Copernicus Books

Sparking Curiosity and Explaining the World

Drawing inspiration from their Renaissance namesake, Copernicus books revolve around scientific curiosity and discovery. Authored by experts from around the world, our books strive to break down barriers and make scientific knowledge more accessible to the public, tackling modern concepts and technologies in a nontechnical and engaging way. Copernicus books are always written with the lay reader in mind, offering introductory forays into different fields to show how the world of science is transforming our daily lives. From astronomy to medicine, business to biology, you will find herein an enriching collection of literature that answers your questions and inspires you to ask even more.

John Naylor

The Riddle
of the Rainbow

From Early Legends and Symbolism to
the Secrets of Light and Colour

 Springer

John Naylor
London, UK

ISSN 2731-8982   ISSN 2731-8990   (electronic)
Copernicus Books
ISBN 978-3-031-23907-6   ISBN 978-3-031-23908-3   (eBook)
https://doi.org/10.1007/978-3-031-23908-3

This Springer imprint is published by the registered company Springer Nature Switzerland AG
The registered company address is: Gewerbestrasse 11, 6330 Cham, Switzerland

The rainbow shines but only in the thought
Of him that looks. Yet not in that alone,
For who makes rainbows by invention?
And many standing round a waterfall
See one bow each, yet not the same to all,
But each a hand's breadth further than the next.
The sun on falling waters writes the text
Which yet is in the eye or in the thought.
It was a hard thing to undo this knot.

Maentwrog[1]

[1] *Hopkins, G. M. (1948), p 128.*

For Billy, lest I forget

The original version of Book Backmatter was revised: index entries for words beginning with B have been included.

Acknowledgements

This history of the rainbow has long been in gestation. Indeed, for several years it was put on hold while I worked on other projects. Over the many years during which I worked intermittently on the manuscript, I owe a huge debt for advice and help I received from Les Cowley, Günther Konnen, Bernard Maitte, Saira Malik, Andrew Pyle, Prof. Simon Baron Cohen, Prof. Frank James, Anthony de Peyer, Prof. Michael Fitzgerald, Prof. John Henry and Claudia Hinz. I also thank William Bradley, Andrew Steele, Roy Bishop, Luke Culver and Colin Leonhardt for the eye-catching photographs they have allowed me to use.

Thanks also to Angela Lahee, my editor at Springer, and the very efficient team that saw the book through the various stages of its production.

Contents

1

Why Rainbows?

… the whole vast sky was stormy and dark. But after a time the westering sun began to shine through the rifts behind us, while before us on the wild flying clouds appeared a rainbow with hues so vivid that we shouted aloud with joy at the sight of such loveliness. For nearly an hour we rode with this vision of glory always before us.[1]

The most unforgettable rainbow I have ever seen was on a wet autumn afternoon many years ago in London while I was on my way home by train. As if on cue, just as we began crossing the Thames, the western sky cleared and a brilliant rainbow appeared in the East, forming a brightly coloured arch that spanned the river a couple of hundred metres downstream. For the ten minutes of the journey, as the train sped southwards through the suburbs, I watched spellbound as the bow kept pace and seemed to leap from building to building, briefly bathing each one in a faint golden light.

I caught further glimpses of the bow in the gaps between buildings after leaving the train and continuing my journey on foot. As I walked up to my front door, I turned for a final look. Yes, it was still there, almost as bright as when I had first caught sight of it almost twenty minutes earlier. I had been followed home by a rainbow!

I remember this spectacle for another reason. At the time I knew, or fancied I knew, a thing or two about rainbows. I had read somewhere that a rainbow is centred on the eye, but it never occurred to me that this meant that one day

---

[1] Hudson, W.H. (1984), p 48.

© The Author(s), under exclusive license to Springer Nature Switzerland AG 2023
J. Naylor, *The Riddle of the Rainbow*, Copernicus Books, https://doi.org/10.1007/978-3-031-23908-3_1

I might see one racing across the sky. A library was clearly not the only place to learn about rainbows. If I was ever to fully understand the descriptions and explanations offered in books I would have to get out and about. It was the start of a quest that has absorbed me on and off for more than a decade and which, I am happy to say, shows no sign of ending.

Ever since, I have made a point of looking for a rainbow whenever the sun shines while it's raining. It's a phenomenon of which I never tire. Like a child anticipating the pleasures of a visit to a toyshop, I rush outside whenever the sky brightens as the rain begins to ease. More often than not I am met by a disappointingly anaemic stub, its muted colours barely discernible against dark storm clouds. Sometimes the bow fades away quickly and at others no bow is visible. But once in a while my persistence is rewarded by one of those huge, bright double rainbows that occasionally dramatically signal the passing of a short-lived shower late in the day. And as if that is not enough, now that I know what to look for, I usually find there's plenty more to see as its brightness and colours alter before my eyes as the rain clears and the sun grows brighter and the raindrops diminish in size.

Although a rainbow is among the most eye-catching atmospheric spectacles that nature has to offer, so distinctive that once seen it is seldom forgotten, it is also among its most elusive. Ask yourself how many rainbows you've seen in, say, the last few months; not many, I'll wager. The fact is, even if you keep your eyes peeled on every occasion when conditions seem favourable, unless you live somewhere where squally showers are common, such as an oceanic island like Hawaii, which Mark Twain once suggested should be renamed the Rainbow Island,[2] a claim recently backed up by meteorological science,[3] you can count yourself fortunate if you catch sight of more than a dozen rainbows in a year, of which only a handful will be memorable.

Yet, since the earliest times, people seem always to have taken more than a passing interest in rainbows. Its multi-coloured arch is such an arresting sight that despite its transient and ephemeral nature, not to mention its rarity—all of which place it on the margins of everyday experience—it has been woven into myths and superstitions the world over, and has attracted the attention of more than its share of the best minds in history. Indeed, since the dawn of secular thought in ancient Greece over two and a half thousand years ago it has been considered the very embodiment of the mysteries of light and colour and leading thinkers in every age have sought to divine its secrets. The sheer amount of intellectual effort expended over the centuries in trying to

---

[2] Twain, M. (1872), p 513.

[3] Businger, S., (2021), p 345-8.

Fig. 1.1 When visible, a double rainbow typically consists of a bright primary bow, a fainter secondary bow, a dark segment between them and a brightening of the sky enclosed by the primary bow[4]

understand the rainbow belies what is otherwise merely another of nature's fleeting and seemingly inconsequential optical phenomena.

Serendipitously, the rainbow proved to be much more than an enchanting, eye-catching phenomenon. It was a laboratory in the sky that informed and illustrated almost all of the earliest attempts to understand the nature of colour and light. But the rainbow proved to be a much more complex phenomenon than it had seemed to the self-styled mechanical philosophers of the seventeenth century who collectively came up with the first satisfactory account of its most conspicuous features because, almost uniquely among naturally occurring optical phenomena, the rainbow exhibits every major property of light, from reflection and refraction to diffraction, interference and polarisation. And in the seventeenth century, only the first two of these properties were understood, if imperfectly (Fig. 1.1).

Regrettably, most elementary accounts of the rainbow begin and end with Isaac Newton, who based his explanation of its size and colours on reflection and refraction alone. Fortunately, stimulated by Newton's explanation, during the eighteenth century, people began to take notice of features of the rainbow that are due to light's other properties, features that Newton failed to address. Yet such was his reputation and authority that a more complete

---

[4] Photo by James Wheeler, used with permission.

understanding of the rainbow was delayed until the early decades of the nineteenth century, when a new generation of natural philosophers—now dubbed
"scientists"—managed by degrees to free themselves from Newton's spell and
delve more deeply into the mysteries of light.[5]

Moreover, until recent times, at several pivotal moments in the history of
optics, the rainbow has been either the proverbial grain of sand that led to
a pearl of understanding or a useful test of a new theory of light. Fittingly,
given its iconic status as one of nature's most celebrated wonders, the rainbow
can also lay claim to being the first natural phenomenon to be explained
using a combination of theory, mathematics and experiment that since the
seventeenth century has become the hallmark of scientific explanation—an
achievement trumpeted at the time as an example of what the new mechanical philosophy could achieve. From a scientific perspective, the fabled crock
of gold at the end of the rainbow turned out to be the theories of light and
colour that it inspired and informed.

By the middle of the nineteenth century, it seemed as if optics had at
last caught up with the rainbow. Light was now considered to be a vibration within a rigid yet invisible medium known as the æther rather than the
stream of material corpuscles envisioned by Newton. This made it possible
to account for a greater range of optical phenomena, including all the then
known features of the rainbow. By the end of the century, however, the
ancient and venerable science of optics, which hitherto had been considered to be a self-contained science on the assumption that light is one of
the elemental forces of nature, had been annexed by the new science of
electromagnetism. But the very experiment that confirmed that light is an
electromagnetic wave—the creation and detection of radio waves in a laboratory in 1886—threw up an apparently trivial phenomenon, which became
known as the photoelectric effect and which sparked the revolution in physics
that took place during the first half of the twentieth century. Photoelectricity
suggested that light has a dual nature as a particle and a wave, something
that eventually blurred the sharp distinction between matter and light and
prepared the ground for quantum theory.

The electromagnetic theory of light made possible a more thorough
account of the rainbow's features, though it did not add all that much to
what had been achieved with the mechanical wave theory of light that it
replaced, at least not for the casual observer. But if the explanation of the
rainbow is now substantially complete, there remain several loose ends, for
no theory of light takes into account of the meteorological circumstances in

---

[5] The term "scientist" was coined by William Whewell (1794–1866) during a meeting of The British
Association for the Advancement of Science on 24 June, 1833. See: Snyder, L.J., (2011), p 3.

which rainbows occur. Due in no small part to its rarity, which makes it difficult to settle unanswered questions such as how to explain bows of variable
curvature that sometimes appear in pairs and are known as twinned bows
and the effect of the size and shape of raindrops on the curvature and the
colours of the rainbow, the rainbow continues to throw up the occasional
intriguing optical puzzle. Though these no longer command much attention
from mainstream science, they still provide sport for a large number of scientists, amateur and professional, prepared to scan the skies and speculate about
the causes of unusual rainbows.

Although science no longer has a use for it, the rainbow retains a place
in the history of ideas because the story of what people have made of the
phenomenon, which has its origins in prehistory, provides us with a series
of snapshots that chart the changes that have occurred in our ideas about
the nature of world, from the anthropomorphic symbolism of the myths of
pre-scientific times to the last gasp of the materialist Newtonian worldview
during the opening decade of the twentieth century and beyond to the era
quantum physics. Furthermore, the close relationship between the rainbow
and ideas about light and colour, whereby developments in the optics led to
a better understanding of the rainbow while the rainbow repaid the favour by
acting as a test bed for those ideas, puts the rainbow at the heart of a wider
history of optics and by extension of science, for speculation concerning the
nature of light has been at the heart of mainstream developments in science
ever since the scientific revolution that took place during the seventeenth
century. Indeed, as we have just noted, at the turn of the twentieth century,
the phenomenon of photoelectricity showed that light holds the key to the
secrets of matter and energy.

But science is a flesh and blood activity, and so its history is not confined
to abstract ideas and impersonal discoveries. It involves individuals whose
successes and failures owe as much to their character and circumstances as it
does to the brilliance of their intellect and the milieu into which they were
born. Moreover, as we shall see in the pages that follow, the history of their
collective endeavour has never been a seamless tale of progress. Not everyone
shared the same ideas or was working towards the same ends. Rivalries,
heated disagreement and bitter personal animosity have always been part and
parcel of that fabled impartial search for truth that we fondly imagine is the
archetypal scientist's only real concern. Nor has the progress of science always
been an unstoppable march to the sunny uplands of truth and certainty, and
it is only with hindsight that we are able to see which ideas have stood the test
of time. At the same time, like all human endeavour, science is an unceasing

work in progress, which is one of the reasons why its history and its discoveries are perennially fascinating, even though its ideas are often difficult to follow for the layperson.

Rainbows are interesting for another reason: it is one of those natural phenomena—of which I believe there are very few—that are of interest to both the poet and the scientist and thus offers an opportunity for them to make common cause. Rainbows appeal to both sides of our nature: the rational (the assumption that natural phenomena have discoverable physical causes) and the emotional (the recognition that natural phenomena can have a meaning that transcends the material world). These viewpoints need not be mutually exclusive. Indeed, they shouldn't be because both the poet and the scientist seek to make sense of experience, though they differ in their methods. The poet seeks an affective engagement with nature; the scientist holds nature at arms' length the better to study it. Unfortunately, this has persuaded all too many people to take sides and assume that scientific explanation robs the rainbow of beauty and meaning, reducing it to just another humdrum event within the natural order. They assume that because scientists make no allowance for subjective experience, their account of the world offers little to nourish the spirit; only the poet can do that. John Keats (1795–1821), the English poet, famously accused science of consigning the rainbow to the "dull catalogue of common things"[6] and of stripping it of wonder. Yet, as you will discover in the pages that follow, the scientist's rainbow turns out to be a far more varied and enthralling phenomenon than the iconic multicoloured arc of the poet's imagination. Arguably, science makes the poet's rainbow shine more brightly if only because it shows us what to look for when we see it, encouraging us to look at the phenomenon more attentively. Moreover, science continues to make entries into Keats' "dull catalogue of common things" under the heading of "rainbows" that would have been unknown and unimagined by him.

In the pages that follow I hope to show that the rainbow is a phenomenon that bridges the supposed chasm between the arts and the sciences. Beginning with the many circumstances in which one can see a rainbow and descriptions of its salient features, this book recounts and explains the myths and superstitions about rainbows and what poets, painters and, above all, scientists have made of the phenomenon and its role in the history of how they came to understand the physical nature of light and colour.

---

[6] Keats, J. (1909), p 41.

2

The Naming of Parts

Rainbow physics is true, but not real, the rainbow we see is real, but not true.[1]

There never seems to have been a time when people were either unaware of rainbows or indifferent to them. Given this enduring interest, it is rather surprising to find that when one asks around it soon becomes apparent that informed knowledge about rainbows is almost as rare as the sight of a rainbow itself. When pressed, most people admit to knowing little more than that it's an arc of many colours and that it has something to do with sunlight and rain. To add to their confusion, the interplay between sunlight and water in form of either liquid droplets or ice crystals gives rise to a veritable zoo of luminous arcs, some of which are as colourful as rainbows. How can one be sure that a particular coloured arc is a rainbow and not, say, a circumzenithal arc, a circumhorizon arc, a parhelion, or a corona, all of which are multicoloured arcs frequently mistaken for rainbows?

If you've never seen these arcs, or know little or nothing about them, it's not much help to be told that circumzenithal arcs and parhelia are seen in clouds composed of ice crystals or that coronas are seen in clouds composed of liquid drops. What you want to know is what they look like and the circumstances in which you are likely to see them. But if all you want is to avoid mistaking them for a rainbow, there is a simple way to do so without the need to know anything about them because they all are seen on the same

---

[1] Berry, M. (2020).

© The Author(s), under exclusive license to Springer Nature Switzerland AG 2023
J. Naylor, *The Riddle of the Rainbow*, Copernicus Books,
https://doi.org/10.1007/978-3-031-23908-3_2

side of the sky as the sun. A rainbow, on the other hand, is always seen on the opposite side of the sky from the sun.[2] Of course, as you may already know, it's only worth looking for a rainbow when it's raining and the rain is directly illuminated by sunlight. Your chances of seeing a rainbow are greatly increased if you are within a few hundred metres of the rain and the sun is close to the horizon, i.e. either early morning or late afternoon.

To see a rainbow, turn your back to the sun and look up at the rain in the direction of your shadow. With luck, you'll see a segment of a luminous, semi-circular arc consisting of several parallel bands of colour. The outermost band is always red and the innermost one either blue or violet. You will usually be able to make out bands of other colours sandwiched between them, such as orange, yellow and green. These colours are, in fact, an imperfect spectrum of sunlight. You may recall from your schooldays that sunlight can be dispersed into a sequence of colours by passing it through a glass prism. Isaac Newton gave the name *spectrum* to this sequence and found that its colours always occur in the same order: red, orange, yellow, green, blue and violet.[3] In the case of rainbows, raindrops act as prisms, but for a number of reasons that we'll go into in a later chapter, raindrops can't match the degree of colour separation or brightness achieved by a prism. As a consequence, the colours in a rainbow are never as vivid as those seen when sunlight passes through a prism.

A larger, broader, fainter arc is often visible outside the first one. It, too, is composed of parallel bands of colour, but their order is reversed: the red band is on the inside of the arc, the violet one on the outside. To distinguish these arcs from one another the inner one is known as the primary bow and the outer one as the secondary bow. Although the primary bow is always visible when you see a rainbow, it is most unlikely that you will see a secondary arc without a primary arc for reasons that will be explained in a later chapter.

Another thing to look out for is that when the secondary bow is visible, the gap between it and the primary bow is often noticeably darker than the sky enclosed by the primary bow. This is usually known, aptly if unimaginatively, as the dark band, and occasionally as Alexander's dark band after Alexander of Aphrodisias, a second century Greek philosopher who appears to have been the first person to draw attention to it. Less frequently, and usually as the rain begins to ease, one or more faint, narrow bands of pastel pink and green are visible inside the primary arc, just beyond the violet band. These are known

---

[2] There is, in fact, another circular multicoloured coloured arc that, given the right conditions, occurs in the same part of the sky as the rainbow. This is the so-called "glory" which can be seen when the sun casts your shadow onto a cloud or a fog.

[3] Guerlac, H. (1965).

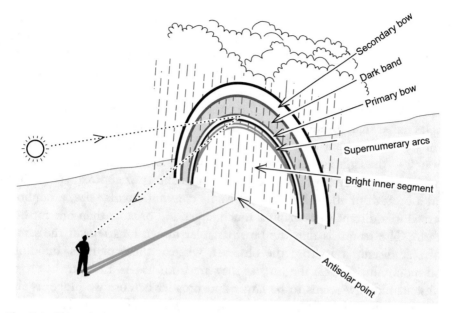

Fig. 2.1 The relative position of observer, sun, primary and secondary bows. Note that a rainbow is not a flat disc as implied in this diagram, it is a three-dimensional cone that stretches into the rain. Nor can you ever see it from the side, as shown here

as supernumerary arcs and are often confined to the apex of the primary bow. On a handful of occasions, extremely faint supernumerary arcs have been seen on the outer edge of the secondary arc (Fig. 2.1).

If you've seen several rainbows, you will know that the intensity of their colours varies enormously from one occasion to the next and that the overall brightness of a bow can change before your eyes. Other things being equal, the brightest rainbows are seen in moderate showers. Atmospheric conditions also play a part: a hazy sky reduces the amount of sunlight reaching the rain. And a rainbow will appear brighter against dark clouds than against bright ones, just as a reflection in a transparent windowpane is easier to see when it's dark on the far side of the glass.

You may also have noticed that rainbows are more common towards the end of the day, when the sun is low in the sky. This is true only where the prevailing rain-bearing weather is from the west, as it is in Western Europe and North America. As the rain clouds drift eastwards, the western sky clears allowing the afternoon sun to illuminate the rain. A morning rainbow seen in clouds blowing in from the west, when the sun is in the eastern sky, is usually short-lived because the clouds soon reach the eastern horizon, blocking the sun. Hence a rainbow seen in the morning is a harbinger of rain, whereas at the end of the day it indicates that the rain clouds are moving away from you. This, of course, is the origin of that well-known weather proverb: 'Rainbow in the morning, shepherd's warning, rainbow in the evening, shepherd's delight'.

One of the rainbow's less obvious features is that its apparent size is constant, no matter how distant the rain in which it is seen. This is why when it comes to measuring the dimensions of a rainbow it is better to use angles rather than distances. Doing so, we find that the outer diameter of the primary bow is about 84°, i.e. just short of a right angle, and its breadth is about 2°—the same as that of your thumb held at arm's length. The secondary bow is significantly larger and has an inner diameter of approximately 104° and a breadth of about 3°. Yet, despite its constant angular size, a rainbow formed in a distant shower looks much larger and broader than one formed nearby. The reason is that the bow's angular dimensions remain the same, however distant it is from the observer, whereas those of trees, buildings and mountains get less the further they are from us. As a result, the arc of a distant rainbow seems to be larger and broader because we judge its size by the apparent size of objects in its vicinity, which, of course, look smaller the further away they are. What we have here is an example of the so-called moon illusion, which you are bound to have noticed, that makes the moon —particularly the full moon—appear noticeably larger when seen close to the horizon than it does when it is high in the sky.[4]

Most of these observational facts have been known since antiquity, but explaining them proved a challenge to some of the sharpest minds in history. The very earliest explanations of the rainbow, which date from around 500 B.C., recognised that it is a reflection of the sun, though the nature of the reflection was not properly understood until the closing years of the fourteenth century, almost two thousand years later. For much of the intervening period it was assumed that a rainbow was a reflection of sunlight by a cloud and the role of raindrops was all but ignored. And it wasn't until almost the middle of the seventeenth century that the rainbow's most distinctive and mysterious feature, the fact that it is a narrow circular arc, was correctly explained by the French mathematician and philosopher, René Descartes.

Despite what you may have been led to believe, the colours of a rainbow are a secondary matter. After all, a coloured spectrum can be brought about in several ways, and is a feature of a good many of the luminous arcs that are seen from time to time in the open air. It's because most people identify the rainbow primarily with its colours that these other luminous arcs

---

[4] Just like the rainbow, the angular size of the moon is constant. The angular diameter of the full moon is ½°. This photo shows that apparent size of the moon is a fraction of that of the arc of a rainbow: https://tinyurl.com/435sje95. Accessed 3/08/2022.

are frequently mistaken for rainbows. In any case, the colours in a rainbow depend on the source of illumination, the intensity of the shower and size of the drops in which the bow is seen. At sunset the arc of a rainbow can be completely red, while a rainbow formed in fog is usually colourless. The real hallmark of a rainbow is its shape, though this is not to deny that if was just a colourless arc we would probably find it less eye-catching than we do.

The key to a rainbow is what happens when sunlight encounters a raindrop. While almost all this light passes straight through the drop, a tiny fraction is reflected and dispersed into its spectral colours within the drop. But if light is reflected from every drop, why doesn't the entire rain shower light up? Why is the reflected light confined to a narrow arc? In fact, the whole shower does brighten. But this is only half the story: the eye also plays a part. Each colour emerges from every drop as a wide cone, which differs slightly in diameter depending on colour. However, as Descartes discovered, the reflected light emerging from a spherical drop is not distributed uniformly, and is most intense at the surface of the cone of light. This is why the drops from which light of the greatest intensity reaches the eye are confined to a narrow circular arc and is the cause of the rainbow's characteristic shape: a narrow semicircular arc. Light from drops that lie outside the arc does not reach the eye, which is why the sky outside the bow appears darker than that within it. At the same time, it isn't possible to see more than one colour from each drop. You have probably noticed how a dewdrop sparkles with different colours in sunlight, and that these colours change when you move your head. It's the same with raindrops: each drop contributes a tiny spot of light of a single colour to the rainbow, so that its arc is really a mosaic made up of a vast number of reflections of the sun separated into its spectral colours by a multitude of drops.

Rain illuminated by sunlight thus teems with an infinite number of potential rainbows that await your eye to pick them out. What's more, since a rainbow is the consequence of a unique angular relationship between the eye, the sun and the drops in which it is seen, the rainbow seen by me can't be seen by you and vice versa. Each of us sees our own rainbow, though it's worth pointing out that the same is true of all atmospheric arcs and much else besides. Ice halos, parhelia, coronas and glories are all centred on the eye of the beholder, as is the glitter path formed when a bright source of light such as the sun is reflected in a large body of ruffled water. The next time you find yourself by water at night, in which several lights are reflected, notice that all the glitter paths appear to converge on you.

The fact that a rainbow is a reflection of the sun means that, contrary to appearances, a rainbow is not actually within the rain in which it is seen. The image you see in a reflection is as far behind the reflecting surface as the object of which it is a reflection is in front, something else you may have learned in school. This means that a rainbow lies as far beyond the rain as the rain is from the sun. The sun is 150 million kilometres from Earth, so a rainbow, an image due to multiple reflections of the sun, must lie 150 million kilometres beyond the Earth. To all intents and purposes, a rainbow might as well be infinitely far away. But just as we can't judge how far away the sun is merely by looking at it because the usual visual clues we rely on to judge depth are absent, we are unable to notice that a rainbow is equally distant, which is why it appears to be within the rain.

Parallax, the apparent change in position of an object against its background that occurs when we change our viewpoint, and which is often the best visual clue we have as to how far a thing is from us, can't be used with a rainbow because it is centred on the eye and thus follows our every move. We can never see a rainbow from the side because the eye is an integral part of the bow, situated at the apex of a cone of light that stretches away from us into the rain. We can't walk around a rainbow as we can, say, a tree. Nor can we see anyone else's rainbow, all of which makes it impossible to avoid the impression that a rainbow is a flat, two-dimensional arc located within the rain in which we see it.

Nevertheless, despite being no more tangible than the reflection that stares back at you from a mirror, a rainbow is real because light from the drops in which it is seen enters your eye. It's not an illusion like, say, the so-called Mach bands that make the inner edge of a fuzzy shadow look darker than it really is and the outer edge lighter, or the coloured shadows that can sometimes be seen at twilight, when the sun is at or just below the horizon while the sky is still bright and blue.[5] These illusions are the result of the way in which the visual system processes neighbouring patches of light and colour, and so do not exist on the surfaces in which we see them.[6] But, like a mirage that can fool you into believing that there is a body of water in the distance, the rainbow before your eyes deceives you into believing that there is an arc in the sky. It's not there, of course: rainbows may be real, but they are not true.

Stripped to essentials, then, a rainbow is a narrow luminous arc of many colours that can be seen when rain is illuminated by sunlight. But this terse description doesn't take into account that no two rainbows are alike or the many circumstances in which rainbows are seen, and so doesn't do justice

---

[5] Naylor, J. (2002), pp 35–39.

[6] You can, nevertheless, photograph Mach bands and coloured shadows because the eye will process the images in the same way as it does the original source.

to their delightful variety. Rainbows are, in fact, variations on a theme of drops of water, sunlight and the eye. For a start, they are not seen exclusively in rain; they can also be seen in mists and fogs, in freshwater and saltwater sprays and even in drops deposited by a fog on spiders' webs or blades of grass. Nor is direct sunlight necessary: rainbows can be formed by sunlight reflected in water, by moonlight and by artificial illumination such as streetlamps or searchlights. As a result, the history of the rainbow is full of reports of rare or unusual rainbows and of rainbows seen in unusual circumstances, all of which makes the rainbow an endlessly fascinating phenomenon.

Even a semicircular bow, the one that comes to mind when we think of rainbows, is visible only when the sun is just above of the horizon and directly illuminates a rain shower that occupies much of the sky opposite the sun. At other times of the day, when the sun is high in the sky, a rainbow is reduced to a mere segment of a semicircle. In fact, the primary arc is no longer visible from the ground when the sun is more than 42° above the horizon. This angle is, of course, equal to the radius of the primary bow and is the basis of a useful rule of thumb, which is that it's only worth looking for a rainbow when the length of your shadow is noticeably greater than your height. This happens when the sun is less than 45° above the horizon. The closer the sun is to the horizon, the longer your shadow and the more of the semicircular arc is visible.

This makes rainbows seasonal because the height to which the sun rises depends on both latitude and season. During summer in London, which is at latitude 50°N, the sun is more than 40° above the horizon from about 8 am to 4 pm, so rainbows can't be seen between those hours, whereas in winter rainbows can in principle be seen at any time of the day.[7]

But don't expect always to see a rainbow, however ideal the conditions may appear to be. As any dedicated rainbow-spotter can tell you, rainbows are more often than not fickle, fleeting apparitions. Rain clouds are always on the move and the sun may be too high in the sky to form a rainbow. By the time you have realised that conditions are ripe for a rainbow, the shower may have retreated beyond the point where you might see a complete arc or, indeed, even a segment of one. On other occasions clouds may prevent sunlight reaching the shower and all you see is a short section of the arc, perhaps a many-coloured shaft bridging the gap between ground and clouds. These stumpy arcs are known as weather galls. And without a clear view across

---

[7] Use Richard Fleet's *GraphDark* application to determine the time of day and the season during which a bow may be visible at your latitude: https://tinyurl.com/4u3m7wud. Accessed 2/02/2022. Given the right conditions of sun and rain, at the latitude of London (51°N) rainbows are visible at any time of the day between the end of September to the end of March.

open country, you may not see a full arc if nearby trees and buildings obscure the foot of the bow, usually its brightest and most vividly coloured section.

The thing to keep in mind is that a rainbow is nothing if it is not perceived; it relies for its existence on been seen, which is why the sight if a rainbow, however fleeting or ordinary, should be an occasion to stop and look. Rainbows seen in unusual circumstances, or which have unusual features are, of course, especially memorable.

Take the seldom-seen moonbow, a rainbow formed by moonlight. Moonlight is, of course, reflected sunlight. But the moon's surface is as dark as asphalt and makes a poor mirror, making a moonbow an extremely pale version of a solar rainbow. At its brightest, when the moon is full, moonlight is about half a million times less bright than sunlight, barely enough to stimulate our colour vision. Most people who have seen a moonbow have usually found that it is colourless and easily mistaken for a cloud until its shape becomes apparent. Indeed, there are only a few days around the time of the full moon when you are likely to see a moonbow.[8] At other times during the lunar cycle the moon isn't bright enough to form a visible bow. The vagaries of weather contribute to its rarity. Local rain showers, those short-lived showers in which rainbows are most often seen, usually die out after sunset. Taking all these things into account, it has been estimated that the conditions necessary for a moonbow occur about a hundred times less often than those for a daylight rainbow.[9] Add to this that a moonbow will only be visible after the end of astronomical twilight when the sky is totally dark, by which time the moon may be too high in the sky for a bow to form or you may be safely tucked up in bed, and you will begin to appreciate that however determined you are, seeing a moonbow may well be a once-in-a-lifetime event.[10]

Colourless bows are also possible in daylight if they are formed in fog or cloud. Climbers and aeroplane passengers are better placed to see this type of bow than someone on the ground because they are more likely to have an unobstructed view of bank of fog or cloud illuminated by the sun. The earliest records of these colourless bows go back to the late thirteenth century, but accurate descriptions of their appearance and the circumstances in which they occur date from the seventeenth century. Edmé Mariotte (1620–1684), a leading French natural philosopher in his day, described seeing such a bow one September morning.

---

[8] Richard Fleet's *DarkGraph* can be used to determine when a moonbow is likely to be visible.

[9] Humphreys, W. J. (1938).

[10] The only moonbows I have seen are those I created in the spray from my garden hose.

There was a dense fog at sunrise. An hour later the fog cleared bit by bit; an east wind swept the fog away some two or three hundred paces from where I was and the sun shone brightly over the fog; I saw a rainbow, which in size, position and shape was similar to the common rainbow. It was completely white though a little darker along its outer edge; its middle portion was very bright, surpassing that of the rest of the fog.[11]

Unlike a rainbow, in which several colours are more or less clearly discernible, a fogbow is usually colourless because its colours overlap, something that occurs when drops are very small. Occasionally, a fogbow sports a dark inner fringe between the main bow and the first supernumerary arc. Although the diameter of a fogbow is always less than that of a rainbow, its arc can be up to three times as broad. Lunar fogbows are also possible and may even be more frequently seen than lunar rainbows because fog is more common at night than short-lived rain showers.[12]

Nor is a daytime rainbow formed in rain always multicoloured because when the sun is at the horizon its light must pass through a far greater amount of air to reach the ground than at other times of the day. In doing so, sunlight is stripped in large measure of its blue and green light, which is why a rising or setting sun appears reddened (Fig. 2.2). As the following account shows, a rainbow formed in these circumstances can sometimes appear wholly red.

There was a sharp shower for about 15 minutes just as the sun was sinking towards a cloudless western horizon. Looking towards the east from a point about 80 ft above Wellfleet Bay [Cape Cod, Massachusetts], I saw an unusually brilliant primary bow at its maximum altitude, with one end in the water and the other end on the shore. As I watched, the blue, green, yellow and orange portions were quickly wiped out, the entire operation taking place in not more than 1 second. There remained a bow of a single colour, red, only slightly less brilliant than before, and in width about a quarter of that of the original bow. I turned to the west and found that the sun had disappeared completely below the horizon, though the familiar red afterglow was still strong. No further change took place in the appearance of the rainbow for perhaps 30 seconds, when it suddenly vanished.[13]

---

[11] Mariotte, E. (1717), p 267.

[12] The earliest published report of a fogbow is often incorrectly credited to Antonio de Ulloa (1716–1795), a Spanish naval officer who accompanied a famous expedition mounted by the Académie Royale des Sciences in 1735 to determine the precise length of on degree of latitude at the equator. Ulloa's bow, as the phenomenon came to be known, was widely discussed at the time without anyone realising that Edmé Mariotte had seen a fogbow some 60 years earlier.

[13] Palmer, F. (1945).

Fig. 2.2 Sunset rainbow in Fairbanks, Alaska shortly before sunset 21 June 2020. The absence of the blue end of the spectrum in the sun's light at sunset is the reason why only the red band of the bow is visible and why the sky enclosed by the primary bow has a rosy hue. *Photo* By Luke Culver, used with permission

A rainbow can also be formed by sunlight reflected by a large body of water, such as a lake, that lies between you and the sun. Such rainbows are known as reflection bows and are rare because, like the moonbow and the fogbow, the conditions necessary for their formation are not often met with. The earliest description of a reflection bow is to be found in a letter written by Edmund Halley, the English astronomer best known for the comet that bears his name, and published in the Philosophical Transactions, the journal of the Royal Society, in May 1698. Halley, wrote that he saw such a bow between 6 and 7 p.m. when he "went to take the air upon the walls of Chester", a town in England. At first he saw only the primary and secondary bows, but a short while later "with these two concentric arches there appeared a third arch, near upon as bright as the secondary iris, but coloured in the same order of the primary, which took its rise from the intersection of the horizon and the primary iris, and went cross the space between the two, and intersected the secondary".[14] He estimated that it lasted about 20 min and concluded that the extra bow was due to sunlight reflected by the waters of the estuary of the River Dee, which lies west of Chester (Fig. 2.3).

In fact, there is no need to wait for rain to see a rainbow because a spraybow can be seen in fountains and waterfalls whenever the sun is shining.

---

[14] Halley, E. (1698).

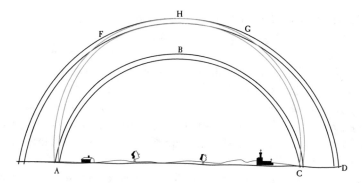

Fig. 2.3 Copy of Halley's sketch of his reflection rainbow.[15] AHC is the rainbow due to reflected sunlight. See Fig. 10.3 for details of how this type of bow comes about

As long as you can get close to the spray, while keeping the sun at your back so that your shadow points in the direction of the spray, you should be able to see a bow, even if it's only a short segment of one, whatever the time of day. Fountains specifically designed to create artificial rainbows were a popular feature in Renaissance gardens and were the inspiration for Descartes' mathematical account of the rainbow.[16] And some waterfalls, such as the Victoria Falls on the border between Zimbabwe and Zambia, are renowned as reliable sources of spectacular spraybows and moonbows.

A spraybow may not be as bright as a rainbow, but what it lacks in brightness, it makes up by being under your control. Spend half an hour on a sunny day with a garden spray and you'll discover several of the rainbow's lesser-known secrets such as that the size and number of drops play a crucial role in determining how bright and colourful a rainbow can be. You can also confirm that the angular size of the arc is constant no matter how far you are from the drops. Step back with your thumb held at arm's length and parallel the arc of the bow: however far you are from the spray, the rainbow arc is always the same width as your thumb. And if you have never seen a moonbow, you can always create a spraybow on a moonlit night. While no substitute for the real thing, you'll find that it has an eerie fascination all of its own.[17]

Stick your nose into the spray and you'll see two slightly overlapping bows, one for each eye; close one of your eyes and its bow vanishes. Here you have the answer to a question you may have asked yourself: is it possible to reach the end of a rainbow? What people probably have in mind when they ask themselves what it would be like to reach the end of the rainbow is that

---

[15] Halley, E. (1698).

[16] Werrett, S. (2001).

[17] It helps to have adjusted your eyes to darkness, a process that takes some 20 min.

they will be immersed in a multi-coloured pool of light. Disappointingly, as experiments with a garden spray reveal, this is not possible because the angular dimensions of the bow remain constant as you approach the spray. Hence its physical dimensions shrink until it finally vanishes just as you get to the curtain of drops. It strikes me that there is a fairy-tale morality about this. If your motive for reaching the end of the bow is to get your hands on that fabled pot of gold, then it is fitting, in a fairy-tale sort of way, that both the rainbow and its treasure should vanish just as you get close enough to touch them.

If you are disheartened by this, you'll be cheered to discover that you can create a circular spraybow. Stand close enough for the shadow of your head to fall within the spray and the resulting bow will be a circle except for a segment at 6 o'clock where your shadow falls on it. Why then, if circular spraybows are possible, are rainbows invariably arcs and not full circles? In fact, circular rainbows are possible, but the necessary conditions are seldom met with. The key is to be able to look down on the rain so that there are drops below your horizon. You can stand near enough to the spray to do this. But to look down on rain you have to be well above ground, something that is possible either by standing on a mountain peak or taking a ride in a helicopter (Fig. 5.8b). Unsurprisingly, skydivers occasionally find themselves heading for a circular bow.[18] What perhaps they may not realise is that because the bow is centred on their eye, they will eventually pass through its centre without having to alter direction (Fig. 2.4). Of course, the bow grows physically smaller as they approach it so it will have vanished by the time they reach the drops in which it is formed.

Spraybows are fun, instructive fun if you're in an enquiring frame of mind, but they are no substitute for the real thing. They can never match the spectacle of a distant rainbow arching high above a sweeping landscape. Spraybows also lack the unpredictability of rainbows, which is a large part of their charm. It goes without saying that the circumstances in which rainbows are seen are largely beyond our control; we can't conjure up a rainbow on a whim as we can a spraybow. Set your heart on seeing a rainbow, and you must be prepared to bide your time, ready at a moment's notice to drop everything and dash outside when conditions seem favourable. Being in the right place, at a moment when conditions favour a rainbow, depends as much on luck as it does on knowledge or dedication.

The descriptions of rainbows in this chapter are entries in what John Keats, denouncing what he considered to be the stunting of imagination brought

---

[18] For a video of sky divers falling towards a circular bow see https://tinyurl.com/256y5ekv. Accessed 1/08/22.

Fig. 2.4 A magnificent circular rainbow seen from a helicopter flying above Lake Argyle, Western Australia. *Photo* By Colin Leonhardt, used with permission

about by Newtonian science, dismissed as "the dull catalogue of common things".[19] But Keats didn't speak for all his kind because an earlier generation of English poets had welcomed Newton's explanation of the rainbow with open arms. They celebrated it as a triumph of human ingenuity that far from inhibiting people's imagination, enlarged it. Indeed, during the first half of the eighteenth century, poetry extolling Newton's account of the rainbow was an important conduit through which his ideas on light and colour reached a wider audience.

And while these poets were sugaring the Newtonian pill for the masses, Newton's heirs, the natural philosophers of the eighteenth century and the physicists of the nineteenth century, relied on Keats' dull catalogue to furnish the raw material with which stoke their creative fires. Indeed, were it not for the keen interest they took in the rainbow, we might know almost nothing of its wonderful variety. Tellingly, there is no record of a poet contributing to the dull catalogue.

---

[19] Keats, J., 1909, p 41.

Humanity's interest in the rainbow, however, predates by several millennia the era of modern science that began in the seventeenth century, for our fascination with the rainbow began well before the earliest civilizations. The evidence is in the frequent references to rainbows in the myths and superstitions that were an important part of the unwritten knowledge that once was handed on by word of mouth from one generation to the next by our remote ancestors.

3

Tales from the Haunted Air

I notice something and seek a reason for it: this means originally, I seek an intention in it, and above all someone who has intentions, a subject, a doer. Every event a deed – formerly one saw intentions in all events, this is our oldest habit.[1]

If the myths and legends of antiquity tell us anything about the ancient world, it is that our distant ancestors didn't consider the natural world to be either inanimate or impersonal, as we do today. Natural events were seen as the actions of gods and spirits possessed of supernatural powers but whose motives were usually all too human. Human appetites and emotions were thus the templates to which people turned when seeking to make sense of natural events. Nature was deemed to act with a purpose that could be properly understood only in terms of its significance to human life. But by cloaking natural events in human form, myths obscured their physical nature, directing attention away from the event itself and prompting people to dwell on what it might signify.

It follows that in those days people took a view of the rainbow that is very different from the one we have today. For them, its features took second place to what it might signify and its physical nature was of little or no interest. The detailed descriptions of rainbows in the previous chapter would have been dismissed with a shrug of the shoulders as a pointless exercise. Little is made in myth or folklore of the secondary bow or of the colours of the primary

---

[1] Nietzsche, F. (1967), p 294.

© The Author(s), under exclusive license to Springer Nature Switzerland AG 2023
J. Naylor, *The Riddle of the Rainbow*, Copernicus Books, https://doi.org/10.1007/978-3-031-23908-3_3

bow. It wasn't that people then were unobservant; on the contrary, survival depended on a keen eye. But what interested them about the rainbow, as with every other natural phenomenon, was what it stood for rather than what it was in itself. A rainbow was considered to be a portent or a symbol, not an inanimate event, and the sight of a rainbow was not an occasion to study its features or delight in its beauty but an opportunity to dwell on the significance of its occurrence.

We can trace one of the sources of the rainbow's significance for ancient people to one of oldest myths: that of a lost paradise. Accounts of a mythical time at the dawn of the human race when people lived in complete harmony with nature are found all around the world. In that golden age, it was said, humans understood the speech of animals and were able to converse with them as readily as one person can talk with another. Moreover, humans were on intimate terms with the gods and could visit them at will in their home in the sky. But due to some transgression, one that varies from one culture to another, either as a result of idleness, an insult to the gods or the failure to perform a particular ritual, the gods put an end to this golden age and mankind was stripped of these powers. Humans became mortal, sexed, obliged to toil for food and at odds with the animal kingdom.

As a consequence, the heavens became inaccessible except under special circumstances. The dead, of course, were granted access to the next world, though this was by its very nature a journey from which there was no corporeal return. In any case, access did not mean automatic right of entry: souls of the wicked could expect to fall by the wayside before reaching the real of the empyrean and find themselves in a ghastly limbo. As for the living, the only way that ordinary folk could communicate with the gods after the fall was through an intermediary such as a shaman.

Shamanism is thought to be man's most ancient spiritual practice and still flourishes in a few nomadic tribes such as the Samoyads of Northwestern Russia. Among these people there is always a tiny number of men and women who not only claim to have healing powers but also to be able to travel between this world and the next. Shamans, therefore, can intercede on behalf of anyone who wishes to communicate with the gods or who wants to ensure safe passage for the souls of their dead relatives to the next world. Access to the heavens, however, requires a symbolic bridge. Usually this is something tangible such as a tree or a stout pole set vertically in the ground that can be climbed by the shaman. But some shamans employ the rainbow to the same effect.[2]

---

[2] Eliade, M. (1964), p 118.

It's not difficult to understand why shamans consider the rainbow to be a symbolic bridge between this world and the next. Of all the luminous atmospheric arcs, the rainbow lends itself most readily to this role because it so obviously bridges the chasm between earth and sky. You may have felt this at the sight of a rainbow, as have I on occasion. Other luminous atmospheric arcs such as ice halos, and coronas are almost always seen high in the sky. They are remote and inaccessible, obviously not of this world, whereas the arc of a rainbow soars into the sky from the ground. And it's a small step from noticing that a rainbow appears to link earth and sky to regarding it as a potential bridge between this world and the next and thus as a means of communicating with the gods. Indeed, the shaman's occasional use of the rainbow as means of reaching the sky is one of the few remaining examples of the rainbow's ancient and widespread role as a symbolic bridge between this world and the next.

The belief that a rainbow is a link between these worlds also explains why it has been widely associated with death and the afterlife. In some cultures it was considered to be a soul-bridge across which the dead must pass to reach the next world. Native Americans considered it to be the road of the dead and Tibetan Buddhists claim that Buddha ascended to heaven on a seven-coloured rainbow. For the Dayaks of Borneo, the rainbow is a boat that conveys the dead to heaven and they make their coffins in the shape of a boat.[3] And from the fact that ordinary mortals don't return having once gained access to the next world, one can see how the rainbow may have acquired the sinister association with death that it has in so many cultures.

At the same time, ascending the rainbow is fraught with danger. While it may be broad for the righteous, it's as sharp as a razor for the wicked. All who cross it must confront demons and monsters and only heroes and initiates such as shamans can overcome these obstacles. Shamans have foreknowledge of these difficulties since they have undergone ritual death and resurrection as part of their initiation; heroes, as one might expect, use brute strength and valour to force their way across.

One of the best known examples of the rainbow as a bridge between worlds is found in Nordic myth in which Bifröst is a bridge between Midgard and Asgard. Bifröst is said to consist of three plaited strands of fire, the combination of which looks like a rainbow to mortals. Midgard is the world inhabited by human beings and lies midway between Asgard, home to the Aesir, a huge, extended family of gods and Niflheim, home of the dead. The Aesir seldom visit Midgard, though they exercise power over the lives of its inhabitants

---

[3] Wessing, R. (2006), p 212.

Fig. 3.1 Ragnarok, Twilight of the gods[5]

and, except for the souls of dead heroes, no mortal from Midgard is allowed to enter Asgard.

Bifröst is guarded by Heimdall, half god, half giant, who possesses formidable powers. He hardly ever sleeps and his senses are so acute that he can see great distances both day and night, and hear grass sprouting in the fields and wool growing on a sheep's back. Heimdall is mute, however, and so is equipped with a horn, named Gjall (or 'shrieker') with which to sound the alarm. The sound of Gjall is so loud that it can be heard throughout the whole of creation.

Asgard is under constant siege by giants determined to steal the apples that the gods are obliged to eat to preserve their immortality. To protect Asgard, the Aesir have surrounded it with a mighty wall. But the wall was badly damaged during one of the many wars between the gods. The price demanded by the mason who repaired it was the sun, the moon and Freya, the goddess who personifies sexual desire. Unwisely, the gods decided to cheat the mason of his due by preventing him from completing the task on time. From that day, it is said, things began to go wrong. Both mortals and giants were seized with anger and hatred and wars broke out, wars that increased in frequency and ferocity, an era known as Ragnarök (Fig. 3.1).[4]

---

[4] *Ragnarok* is better known by its German translation: *Götterdämerung* (Twilight of the Gods).
[5] Image by Friedrich Wilhelm Heine, Wikimedia Commons, https://tinyurl.com/2vdx62d8.

Ragnarök marks the end of the present cycle of creation. Eventually the giants will manage to force their way past Heimdall and swarm across Bifröst, destroying it as they do so. The ensuing battle between the Aesir and the giants ends in the destruction of everything except for Yggdrasil, the giant ash tree that is the hub and support of the universe, and the two mortals and the few animals that will have taken shelter in its branches. This couple will repopulate Midgard and a peaceful new age devoid of giants and demons will dawn. As we shall shortly see, the theme of a world repopulated by a solitary man and woman, the only survivors of a catastrophe in which mankind had been wiped from the face of the earth, is widespread.

A slightly different version of the rainbow as a celestial bridge is found in the ancient Japanese account of creation. Two gods, the female Izanami and the male Izamagi, are ordered by the other gods to create the world from the primordial ocean. They stand together on the Heavenly Floating Bridge, which takes the form of a rainbow arching over the world, while Izamagi stirs the ocean with the Heavenly Jewelled Spear. When he lifts the spear clear of the water, a drop of water falls from its point and forms the first island of the Japanese archipelago.

In a Hawaiian legend, Kaha'i, the god of lightning, ascends a rainbow in search of his father, Hema, whose eyes have been gouged out by fishermen who used them as bait. Among Hawaiians, the eye is the embodiment of knowledge, which is why Kaha'i wants above all to restore his father's sight.

And in another corner of the globe, in the Forrest River district of Australia, a medicine man must undergo a symbolic death and resurrection during his initiation, a necessary rite of passage for all shamans. First, he is shrunk to the size of an infant by his master and placed in a bag. The master takes the bag into the sky by climbing the Rainbow Serpent, represented by a rainbow. When he reaches the top of the rainbow, the candidate is symbolically killed by being thrown into the sky. The master then inserts tiny rainbow serpents into the initiate's body before bringing him back to earth via the rainbow, returning him to his proper size and waking him.[6]

A rainbow also crops up in the biblical story of Noah and the Flood, though as a sign of divine communication, not as a symbolic bridge. Dismayed and angered by the wickedness of Adam's descendants, God resolves to wipe the slate clean by inundating the earth and drowning everyone and everything. Everyone, that is, except Noah and his family, and everything, except breeding pairs of every living creature. All are to ride out

---

[6] Eliade, M. (1964), p 131.

the deluge in a huge ark built by Noah and his sons according to God's instructions.

When the waters finally abate, and the ark's occupants are once again able to walk on dry land, God promises Noah that never again will He inflict such a devastating punishment on mankind. The sign of this undertaking is to be the rainbow.[7]

> And it shall come to pass, when I bring a cloud over the earth, that the bow shall be seen in the cloud, and I will remember my covenant, which is between me and you and every living creature of all flesh; and the waters shall no more become a flood to destroy all flesh...[8]

To Christians in more pious times than ours this passage suggested that there were no rainbows before the Flood: Noah, not Adam, was the first man to see a rainbow. By and large, however, biblical scholars have taken the view that God gave special significance to a phenomenon that pre-dated the Flood. Nevertheless, uncertainty about the exact role of the rainbow in the biblical account of the flood is apparent in paintings that depict the Flood in which a rainbow is shown above the ark, whereas there is no indication in the Bible of a rainbow appearing during the Flood or, indeed, after it had abated. It was only when he had disembarked from the ark that Noah learned that the rainbow was to be a reminder to God of his promise never again to flood the world.

In his commentary on the significance of this event, The Venerable Bede, a seventh century Anglo-Saxon Benedictine monk and biblical scholar, claimed that blue of the rainbow is a reminder of that covenant and that the red was a forewarning of the future flames of the day of Judgement. "And not vainly does it shine with the colour blue at the same time as red, because blue is the colour of flowing water, red of the future flames, which are placed as a testimony for us." Hence the rainbow is a reminder to every Christian both of God's mercy and the need for individual repentance.[9]

During the Middle Ages in Germany this gave rise to a belief that as long as rainbows are seen following rainstorms there was nothing to fear because one of the signs that the Day of Judgement was at hand was supposed to be that for forty years before the end of the world there would be no rainbows.

---

[7] The biblical rainbow is John Keats' much quoted 'awful rainbow' in his poem, Lamia, part 2, line 231.

[8] RSV Bible (1952), Genesis, Ch 9, 12–15,

[9] Anlezark, D. (2013), p 82–3.

In recent times, several biblical scholars claim that the choice of the rainbow in Genesis as a sign from God to man is based on the fact that the Hebrew word for bow—*qeset*—refers both to a war bow and a rainbow. In fact, there is no specific word in ancient Hebrew for rainbow. Hence the sight of a rainbow in the sky might mean that God has hung up his war bow to show that he is no longer angry with mankind. Furthermore, claim these scholars, there is a precedent for identifying the rainbow as a war bow in the *Enuma Elish*, the Babylonian story of creation, which predates the compilation of Genesis by at least a thousand years.

Not everyone agrees with this interpretation.[10] As we shall see presently, in the *Enuma Elish*, the rainbow is employed as a weapon in a battle between the gods that is then hung among the stars as a mark of victory; it celebrates triumph, not reconciliation. An alternative possibility is that the Genesis rainbow represents an arch that holds back the waters of the heavens. In support of this view, consider the last sentence of God's explanation of the significance of the rainbow: its appearance in the sky means that "… the waters shall no more become a flood to destroy all flesh. The rainbow shall be in the cloud, and I will look on it to remember the everlasting covenant."[11] The only other reference to the rainbow in the Old Testament describes it not as a weapon but as an arch above the dome of heaven. The prophet Ezekiel describes a vision of God surrounded by an angelic host: "And above the dome over their heads there was the likeness of a throne, in appearance like sapphire … Like the bow (qeset) in a cloud on a rainy day, such was the appearance of the splendour all around. This was the appearance of the likeness of the glory of the Lord"[12] Had the ancient Hebrews associated the rainbow with a war bow, Ezekiel's description would have made no sense to them.

There is only one other unambiguous reference to the rainbow in the Bible. In The Revelation to St John, the last book of the New Testament, where St. John describes his vision of the day of judgement and in which he sees Jesus surrounded by a rainbow: "…and lo, a throne stood in heaven, with one seated on the throne! … and round the throne there was a rainbow that looked like an emerald."[13] In most Medieval paintings of Christ in Judgement, his throne is often depicted as a semicircular rainbow, thus drawing on the significance of the Genesis rainbow with its promise of mercy. In these paintings Christ is seated at the apex of the bow, sometimes with his

---

[10] Turner, L.A. (1993).

[11] RSV Bible (1952), Genesis, 9, xv.

[12] RSV Bible (1952), Ezekiel: 1, xxvi & xxviii.

[13] RVS Bible (1952), Revelation 4, ii-iii.

feet resting on a smaller second bow.[14] But other depictions of the event are perhaps truer to the description in Revelation and show Christ is surrounded either by a mandorla or a circular rainbow.[15]

According to a chronology of biblical events published in 1650 by James Ussher (1581–1656), Archbishop of Armagh, the Flood occurred in 2348 B.C. and lasted almost a year. These days Ussher is remembered for fixing the date of creation at midnight on 23 October, 4004 B.C. His name often crops up in historical accounts of the development of geology, where he is usually and unfairly dismissed as a reactionary duffer. But the biblical account of creation was widely believed to be correct until the middle of the nineteenth century, with the Flood being the event responsible for shaping the world's geography and geology.[16] In his day, Ussher was a renowned and respected scholar, and no one took greater care to establish an accurate biblical chronology by, among other things, matching the dates of events in the Bible with those of known historical ones. Ussher's chronology was by no means the only one available to Christians, but it became widely known and accepted throughout the English-speaking world because it was usually printed alongside the relevant verses in early editions of the King James version of the Bible.[17] Ussher's greatest mistake in compiling his chronology was to assume that the Bible is a reliable record of the history of the world. But, of course, he wasn't (and he isn't) alone in taking the Bible at face value.

Tradition has it that the author of the first five books of the Bible, which begins with Genesis and ends with Deuteronomy, collectively known as the Pentateuch, was Moses. But we now know that the Pentateuch was really the work of several hands, a scissors and paste job carried out by unknown scribes sometime between 950 and 540 B.C. and based on at least four distinct sources. The account of the Flood draws on two of these, of which the one that mentions the rainbow is the more recent.

In fact, the account of the Flood in Genesis is almost identical to a far more ancient account of a universal flood that is included almost as an afterthought towards the end of a Babylonian saga known as the *Epic of Gilgamesh*. The story has been gradually pieced together by scholars over several decades from hundreds of clay tablets recovered in 1850 by an English archaeologist and his

---

[14] Single rainbow in Hans Memlin (1430 1494), Last Judgement, Triptych, Muzeum Pomorskie Gdánsk Poland (https://tinyurl.com/2p9ab4s3. Accessed 2/03/22. Double rainbow in Stefan Lochner (1400–1452), Jüngstes Gericht, Wallraf-Richartz-Museum, Colone, Germany (https://tinyurl.com/2ky twej6. Accessed 2/03/22.

[15] Giotto di Bondone (1267–1337): *The Last Judgement* in the Cappella Scrovegni (Arena Chapel), Padua. (https://tinyurl.com/3xvhypw7. Accessed 2/03/22.

[16] Cohn, N. (1996), especially Chaps. 9 and 10.

[17] Gould, S. J. (1993), p 181–93.

Assyrian assistant from the ruins of the royal library of the last great king of Assyria, Ashurbanipal (668–627 B.C.), at Nineveh, in what is now northern Iraq.[18]

Like Noah, Gilgamesh is a legendary figure. He was said to be one of the kings of Uruk, an ancient Sumerian kingdom that flourished in southern Mesopotamia some four thousand years ago and from which modern Iraq takes its name. In the epic that bears his name, Gilgamesh is portrayed as headstrong and wanton, a thoroughly disagreeable character. In answer to appeals from his long-suffering subjects, the gods create Enkidu, a hairy giant, more beast than man, who will be more than a match for Gilgamesh. But before this feral brute can fulfil his allotted task, he is memorably seduced by a prostitute who instructs in the ways of men. In the event, when Gilgamesh and Enkidu first encounter one another, they fight one another but soon after become firm friends and set off on a series of adventures. Eventually their exploits, which include slaying Humbaba, guardian of the gods' forest of Cedar trees, as well as killing the Bull of Heaven, incur the gods' displeasure and they decide that Enkidu should be punished. He falls ill and dies, an event that causes Gilgamesh to confront his own mortality.

Desperate to avoid death, he seeks out Uta-napishti, the only mortal to have been granted immortality by the gods. Uta-napishti tells Gilgamesh how Enlil, the storm god, planned a great flood that would wipe out mankind, and how Ea, wisest of all the gods, had betrayed Enlil by secretly urging Uta-napishti to build a huge vessel and stock it with animals and artisans from each of the crafts necessary to civilization. When the flood abated, the vessel came to rest at the summit of Mount Nimush. Enlil was enraged when he found out that there were survivors but was persuaded by Ea to accept that the decision to wipe out all mortal beings had been a terrible mistake.

Ea berated Enlil for inflicting a deluge on mankind without consulting the other gods. He admonished Enlil that only the guilty deserve punishment.

> On him who transgresses, inflict his crime!
> On him who does wrong, inflict his wrongdoing!.[19]

Here, you may feel, the polytheistic authors of Gilgamesh display a more developed sense of justice than the monotheistic Jews, whose own god, Yahweh, set about dispatching the entire human race without a single qualm. Discomforted by Ea's words, Enlil granted Uta-napishti and his wife immortality, not because he believed they deserve it, but to make amends for his

---

[18] George, A. (1999), p xxii–xxiv.
[19] George, A. (1999), p 95.

attempt to destroy mankind. He then banished them to the ends of the earth, which is where Gilgamesh has had to travel to find them. The epic ends with Uta-napishti convincing Gilgamesh to accept his mortality and return to Uruk.

Uta-napishti doesn't explain why the Enlil decided to destroy mankind. But the reason the god did so is given in an earlier account of a flood visited on mankind by the gods, which predates the earliest known versions of the Gilgamesh epic by at least a thousand years. This is the story of Atra-hasis, an earlier incarnation of Uta-napishti. The older story relates how long-ago mankind became so numerous and noisy that Enlil decided to reduce their number by visiting a series of disasters on them. First came a drought, then a plague and finally a famine. Each time the survivors quickly repopulated the earth and restored the status quo. Exasperated, Enlil resorted to flooding the land so that everyone would to be drowned. Although all the gods, including Ea, agreed to this drastic act, Ea secretly urged Atra-hasis to build a huge vessel and fill it with living creatures and artisans, which is how there were survivors of the deluge.[20] As in The Epic of Gilgamesh, we learn that Enlil is angered when he finds the survivors, but is persuaded by Ea that the flood had been a mistake. Unlike Uta-napishti, however, Atra-hasis is not granted immortality. On the contrary: before the flood, men and women did not die of natural causes, which is why they were so numerous. Henceforth, Enlil decrees that death is to be an inescapable fact of life for all mankind.

There are too many parallels between the biblical and Mesopotamian accounts of the flood for the similarities to be coincidental. In both stories the flood is punishment for man's misdemeanours; the source of the floodwater is rain; the hero is told to construct a huge vessel and fill it with animals; when the flood finally subsides, the vessel comes to rest on a mountain summit. And in both stories the survivors release birds to determine if the flood is at an end and, mollified by the burnt sacrifices of the survivors after the flood, the gods grant them concessions (immortality for Uta-napishti, survival for Noah's descendents).

Futhermore, since the story of Uta-napishti and Atra-hasis pre-dates that of Noah by some two thousand years, there can be no doubt about the origin of the biblical account of the flood. In 587 B.C. Nebuchadnezzar, the king of Babylon, destroyed Jerusalem and exiled most of its inhabitants to Babylon. There the Jews would have learned the story of the flood, always supposing they hadn't come across it before their exile. They remained in captivity for

---

[20] Finkel, I. (2014).

some fifty years until Cyrus the Great, founder of the Persian Achaemenid Empire, conquered Babylon in 539 BC and freed its captive Jews.

There are, inevitably, differences between Mesopotamian and Jewish versions of the flood story because they serve different ends. The biblical account presents the flood as the act of a single supreme being angered by human wickedness, not as the result of a squabble among pagan gods irritated by an ever-increasing number of clamorous mortals. Then, the biblical flood is on an altogether more epic scale: it rains continuously for 40 days and nights and vast quantities of water well up from beneath the earth. In the Mesopotamian version the rain lasts a mere seven days, demonstrating the greater power of the Hebrew god compared to the gods of the Mesopotamians. But the olive leaf that the dove brings back to Noah is a detail that is absent in the Mesopotamian version, since olive trees don't grow in Mesopotamia whereas they do in Palestine. Nor is there any mention of a rainbow in any of the Mesopotamian versions of the flood.

At any rate, there is no overt reference to a rainbow. However, it has been suggested that in the verses that describe the joy with which the gods greet the sacrifice made by the flood survivors, the event is marked with a sign reminiscent of the Genesis rainbow.[21]

Then at once Belet-ili arrived,
She lifted the flies of lapis lazuli that Anu had made for their courtship:
'O gods, let these great beads in this necklace of mine
make me remember these days, and never forget them!'.[22]

But the idea that a necklace of monochrome lapis lazuli stands for a rainbow does seem, on the face of it, to be stretching things. Moreover, the necklace is simply held aloft, it is not placed in the sky to act as a universal sign to mankind; at best the goddess' action seems to be a momentary, private gesture.

The absence of a symbolic rainbow in Gilgamesh is an unexpected omission, for, as we shall see, the rainbow had a symbolic role in the Babylonian scheme of things. It's not as if Babylonians, to mention but one of the many civilisations that flourished in Mesopotamia, weren't interested in atmospheric phenomena, or that they didn't consider them to be portents. As we have noted, to the pre-scientific mind all meteorological and celestial events have supernatural significance. Babylonians faithfully recorded all manner of

---

[21] Lee, R. L., Fraser, A. B. (2001), p 6.
[22] George, A. (1999), p 94.

atmospheric phenomena including ice halos, and rainbows, along with celestial ones such as comets (including the one that was identified by Edmund Halley in the seventeenth century and which now bears his name), meteors, eclipses and, unsurprisingly, stars and planets. However, their interest was, in the main, pragmatic. They considered unusual events to be omens that could be used to divine the gods' intentions. But Babylonians weren't fatalists: if an omen could be interpreted correctly, the gods could be suitably propitiated and trouble avoided.

The rainbow crops up the *Enuma Elish*,[23] the Babylonian account of creation, in the form of a war bow. The epic relates how the restlessness of the younger gods, grandchildren of the primeval demiurges, Apsu (associated with the male sweet waters) and Tiamat (the female salt waters, personified as a dragon in reliefs and seals), so annoys Apsu that he decides to kill them all. When the young gods learn of Apsu's intentions, one of their number, Ea, casts a spell over him and slays him. Tiamat is persuaded by the older gods to seek revenge, prompting Ea to cast a spell over her as well, but it fails to work. It is Ea's son, the mighty warrior god Marduk, who succeeds in killing Tiamat, having been equipped with every possible aid to victory, including arrows of lightning, a war bow in the form of a rainbow and a net held open by the four winds.[24] Marduk splits Tiamat's watery corpse in two, with one half of which he creates the heavens and the other the land. To mark Marduk's victory, Anu, the sky god, hangs Marduk's war bow among the stars.

> Anu raised (the bow) and spoke in the assembly of gods,
> He kissed the bow. "May she go far!"
> He gave to the bow her names, saying,
> "May Long and Far be the first, and Victorious the second;
> Her third name shall be Bowstar, for she shall shine in the sky."
> He fixed her position among the gods her companions.
> When Anu had decreed the destiny of the bow,
> He set down her royal throne. "You are the highest of the gods!".[25]

Marduk was the Babylonian's chief god, so it is hardly surprising that according to the account of his deeds in the *Enuma Elish* we learn that Marduk's price for defeating Tiamat is that he be made first among equals in

---

[23] The title *Enuma Elish* is taken from the first words of the epic: "When on high...".

[24] James, E. O. (1960), p 209.

[25] Dalley, S. (1989), p 263.

perpetuity, "king of the gods of heaven and earth", a god worthy of worship by all.[26]

The association of the rainbow with a war bow was not confined to Babylon. Marduk has a counterpart in Indra the warrior god in Hindu myth. Like Marduk, Indra is a warrior-god chosen by the other gods to destroy the demon serpent Vritra who presides over the cosmic waters and prevents their use and thus stops the world coming into being. Armed with a war-bow Indra kills Vritra with arrows of lightning.[27] The Sanskrit word for rainbow is *Indradhanush*, or Indra's bow. It is more than likely that the story of Indra's destruction of Vitra is a version of the Babylonian myth, with Tiamat replaced by Vitra.

Strangely, according to the Bible the rainbow is not intended primarily as sign to mankind: it's God's reminder to Himself of his covenant with Noah that He has promised to restrain his fury at human wickedness, a memorable example of anger management you might think. Though why God should need to remind Himself of His promise is not explained. But, given that He has decided to do so, a rainbow is the obvious choice, and for two reasons, one natural and the other symbolic. A rainbow, if it appears at all, usually does so as the rain clears away and the sun breaks through the clouds. It's a natural sign that the shower is abating and it's no longer necessary to take shelter. At the same time, it draws on the ancient belief that the rainbow is a link between this world and the next, and thus a sign of divine action of some sort. But it is unlikely that the authors of Genesis would have come up with such symbolism unaided. They almost certainly relied on a pre-existing association between the rainbow and divine communication, though the immediate source on which they drew is unknown to us.

Evidence that the rainbow was considered to be a sign of divine communication well before the compilation of Genesis is found in Greek myths. Here we find that Iris is both the name of a messenger employed by the gods of the ancient Greeks and Romans and the word for rainbow in Greek and Latin. According to the Greek poet, Hesiod (c.700 BC), the author of *The Theogony*, the first systematic genealogy of the gods of ancient Greece, Iris was the daughter of the Titan Thaumas and the Nymph Electra.[28] Her sisters were wind-spirits called Harpies, creatures described in the *Aeneid*, the epic poem by the Roman poet Virgil (70 BC–19 BC), as birds

---

[26] For a fuller account of the story of Marduk, see Cohn, N. (1995), pp 45–9.

[27] Cohn, N. (1995), p 62–5.

[28] Hesiod (1913), lines 265–69.

Fig. 3.2 Iris, messenger of the gods[31]

> With virgin faces, but with wombs obscene,
> Foul paunches, and with ordure still unclean;
> With claws for hands, and looks for ever lean.[29]

Iris, on the other hand, was usually portrayed as a young woman with wings, and her attributes were the herald's staff and a vase, in which, according to Hesiod, she fetched water from the River Styx whenever the gods had to take a solemn oath. The water would render unconscious for an entire year any god or goddess who had lied (Fig. 3.2).[30]

In the Iliad, Homer's epic of the last days of the Trojan War, Iris usually acts on behalf of Zeus, the supreme Olympian god, who uses her to communicate with mortals and, occasionally, with other gods. Early on in the Iliad, Iris is dispatched to warn the Trojans of the approaching Greek army. Later she is sent to advise Hector, son of the King of Troy, that he should not attack the Greeks until their leader, Agamemnon, has been wounded. When Zeus discovers that his younger brother, Poseidon, the god of earthquakes and of the sea, has been helping the Greeks against the Trojans, he sends Iris to command Poseidon to quit the battle or suffer the consequences of Zeus' displeasure.

Iris is also employed by Hera, Zeus's wife, to tell Achilles, the great Greek hero, that Hector intends to seize and defile the dead body of Patroclus,

---

[29] Virgil (1697), Bk III: 217–19.

[30] Hesiod (1913), lines 775–806.

[31] Image by Jean-Édouard Dargent, Wikimedia Commons, https://tinyurl.com/2y5x98p9.

Achilles' greatest friend. Iris urges Achilles to recover the body before Hector finds it. On a couple of occasions Iris acts on her own account. She tells Helen that Paris, the Trojan prince who caused the war by abducting Helen, and Menelaus, Helen's husband, are to fight a duel over her. On the second occasion Iris saves the wounded Aphrodite, the goddess of love and beauty. Iris also helps mortals: she comes to the aid of Achilles, and conveys his prayers to north wind, Boreas, and the west wind, Zephyr, imploring them to fan the flames of Patroclus' funeral pyre.

Curiously, given the association between Iris and the rainbow, Homer never mentions the rainbow when describing occasions when Iris appears. There are only two direct references to the rainbow as a sign of divine communication in the Iliad. Homer describes the appearance of the decorative snakes on the breastplate worn by Atreides, one of the Greek warriors:

> On either side three snakes [on the breastplate] rose up in the coils towards the opening for the neck. Their iridescent enamel made them look like the rainbow that the son of Chronos [i.e. Zeus] hangs on a cloud as a portent to mankind below.[32]

Elsewhere, Zeus sends Athene, the goddess of wisdom and daughter of Zeus, to rally the Greeks.

> Wrapping herself in a lurid mist, like some sombre rainbow hung in the sky by Zeus to warn mankind of war or the coming of a cold squall that stops work in the fields and brings discomfort to the sheep, she dropped among the Greek soldiery and put fresh heart in one and all.[33]

Nevertheless, despite the fact the Homer never directly associated Iris with the rainbow, both the Greeks and the Romans linked iris the rainbow with Iris the goddess. In our best source of the classical myths of the ancient Graeco-Roman world, *Metamorphoses* by the Roman poet Ovid (BC–17 AD), Iris is explicitly identified with the rainbow on a number of occasions.

Metamorphoses begins with the creation of the world and of the first humans. In terms that echo those found in the Epic of Gilgamesh and in Genesis, Ovid describes how the actions of the first humans so angered Jupiter, the ruler of the gods, that he decided to wipe them out with a flood.[34]

---

[32] Homer (1950), XI, 26–28, p 197.
[33] Homer (1950), XVII, 547–50, p 330.
[34] Jupiter and Juno are the Roman versions of Zeus and Hera.

The thunder crashed and storms of blinding rain poured down from heaven. Iris, great Juno's envoy, rainbow-clad, gathered the waters and refilled the clouds.[35]

At the end of the penultimate book of *Metamorphoses*, Ovid tells of the deification of Romulus, one of the legendary founders of Rome. Juno, wife of Jupiter, moved by the grief of Romulus' wife, Hersilia, sends Iris to her with instructions about how she can herself be deified so that she can join Romulus.

His wife, Hersilia, was mourning him as lost, when royal Juno ordered Iris to descend to her, by her rainbow path, and carry [Juno's] commands, to the widowed queen…[36]

But the clearest link between Iris and the rainbow's coloured arc is found in the *Aeneid*, which takes up the story of Aeneas, one of the *Illiad*'s Trojan heroes, after the fall of Troy. Virgil's Aeneas is urged by the ghost of Hector to escape before the Greeks capture and sack Troy and travel to Italy, there to found a new city—a city that Virgil intends his readers to associate with Rome. But throughout the subsequent odyssey, Juno, the Roman equivalent to Hera, who in the *Illiad* supports the Greeks against the Trojans, tries to hinder him again and again. As he and his companions approach Sicily, Juno persuades Aeolus, the god of the winds, to create a great storm that forces them to turn about and head in the opposite direction towards Carthage. Juno hopes that Aeneas will fall in love with Dido, the queen of Carthage. Dido becomes infatuated with Aeneas, and they briefly become lovers. But eventually he heeds the call of duty, and sets off once more for Italy leaving Dido heartbroken. Inconsolable, she kills herself. But having died by her own hand, her soul cannot quit her body. Juno takes pity on her plight and commands Iris to release her from her body.

So dew-wet Iris flew down through the sky, on saffron wings, trailing a thousand shifting colours across the sun.[37]

Later, when Aeneas has reached in Sicily, Juno attempts once again to thwart his plans to reach Italy by sending Iris to set fire to his ships while Aeneas and

---

35 Ovid (1986), Book I: 270.
36 Ovid, (1986), Book XIV: 829.
37 Virgil (1697), Book IV, 700–1.

his comrades are marking the anniversary of his father's death with funeral games.

> Here Fortune first alters, switching loyalties. While they
> with their various games, are paying due honours to the tomb
> Saturnian Juno sends Iris down from the sky to the Trojan fleet
> breathing out a breeze for her passage, thinking deeply
> about her ancient grievance which is yet unsatisfied.
> Iris, hurrying on her way along a rainbow's thousand colours
> speeds swiftly down her track, a girl unseen.[38]

A very different idea of the rainbow is found across a huge swathe of the tropics, from the mountains and jungles of South America to those of equatorial Africa and South-Eastern Asia. In all these places the rainbow is associated with a gigantic, malevolent serpent that usually bodes ill for anyone who goes near it.

Such beliefs were still current as recently as 1953, as an anthropologist visiting the Semai people of Northern Malaya discovered. He was taken aback by the obvious apprehension with which a young woman was watching a rainbow. He was told that the Semai considered the rainbow to be a gigantic, noxious serpent that lives in rivers, from where it ascends into the sky; anyone walking under the rainbow would fall prey to a fatal fever. For the Semai a rainbow is "the shadow which arises from the body of a great snake, which lives in the earth. The red of the rainbow is its body, the green its liver, and the yellow its stomach". Another Malayan tribe, the Semang, whose word for rainbow means serpent, believed that rain that falls when a rainbow is visible is extremely dangerous. Unless protected by a suitable amulet, anyone caught in this rain is likely to become ill. The ground where a rainbow touches the earth's surface is considered dangerous and it is unhealthy to reside anywhere near such a place.[39] Similar ideas are found in Java, where it is also believed that the rainbow is a serpent that drinks from the sea and vomits this water over the land as rain.[40] At the same time the Javanese rainbow represents a boat rowed by an osprey between heaven and earth. In fact, the Dayaks of Borneo, who share this view of the rainbow, make their coffins in the shape of boats in anticipation of their final journey.[41]

---

[38] Virgil (1697), Book V, 604–14.
[39] Loewenstein, J. (1961).
[40] Hooykaas, J. (1956), p 306.
[41] Wessing, R. (2006), p 212.

Throughout tropical West and East Africa, where snake-worship is extremely common, the rainbow is associated with the python. The rainbow is said to be a celestial serpent that emerges from its hiding place when it rains. Pygmies of the Eastern Congo have a particular dread of the rainbow, which they regard as a gruesome snake-monster that devours human beings and brings about disasters.

A similar view is found among Amazonian tribes. The Botocudo of Eastern Brazil believe that a great snake is the lord of the waters: it signals to the rain and makes it fall. They call the rain that falls when a rainbow is visible "the urine of the great snake" . And in the Peruvian Andes, to this day, the rainbow is considered to be the body of a giant double-headed serpent, which the natives call *Amaru*. As a rainbow, *Amaru* is said to rise out of one underground spring, arch across the sky and bury its second head in another spring.[42] The link between the rainbow and serpents can be traced to the pre-colonial Incas, who believed that they were descended from *Amaru* and considered the rainbow a portent of both good and evil depending on the circumstances in which one appeared.[43] The Inca deity of the rainbow was called *Cuichu* (Fig. 3.3).

In his history of Peru, Bernabé Cobo (1580–1657), an early Spanish chronicler of the of Peru, claimed that a rainbow flag represented the Inca himself: "the sign of the Incas was the rainbow and two parallel snakes along the width with the tassel as a crown"[44] The motif was occasionally incorporated in the coat of arms of Spanish-American patricians during colonial times. There are no giant serpents in the Peruvian highlands, however, so it is likely that the Incas acquired the association between serpents and rainbows from the lowland tribes that inhabited the Amazon forests that border the eastern foothills of the Andes. One of the tributes demanded by the Incas of their forest-dwelling subjects were giant snakes such as Boas and Anacondas. And with those writhing, glistening serpentine tributes, we must surmise, came the myth of the rainbow serpent.

Further evidence of the association between serpents and rainbows among the Amazonian tribes was unearthed by Claude Lévi-Strauss (1908–2009), the Belgian anthropologist who made a detailed study of Amazonian myths while he was living in Brazil between the two World Wars. He found that the rainbow was widely believed to be malevolent snake-spirit associated with

[42] Urton, G. (1981), p 115.
[43] MacCormack, S. (1988), pp 1001–5.
[44] Cobo y Peralta, B. (1893).

Fig. 3.3 The Inca's personal coat of arms in which two open-mouthed serpents are linked by a rainbow

death and disease, which in turn is believed to be the result of poison—for without poison to make them ill men would live forever.[45] Another element of this myth is that birds acquired their coloured plumage when they killed the rainbow snake and divided its skin among themselves. Lévi-Strauss claimed that as a consequence of the association of polychromism with a malevolent rainbow snake, there was a marked avoidance of multicoloured decoration among the Bororo, a tribe who inhabit northern Paraguay.[46]

There is also an astronomical dimension to these rainbow beliefs that involves the rainbow's colours. It seems that both the lowland and highland tribes of Latin America believe that the rainbow serpent appears during the day as a rainbow and at night as an elongated dark patch in the Milky Way west of the zodiacal constellation of Scorpio. The lack of light and colour in

[45] Lévi-Strauss, C. (1970), p 246.
[46] Lévi-Strauss, C. (1970), p 321.

the Milky Way is regarded as the celestial counterpart of the bright, many-coloured rainbow.[47] The Incas, who flourished from the 13th to the early 16th centuries, had a more elaborate version of this belief for they associated this dark patch, which they called Mach'Acuay, their word for snake, with the yearly cycle of dry and rainy weather, for the earliest sighting of this dark constellation in the Western night sky in August coincides with the onset of the rainy season, which is when snakes begin to hatch.[48]

Another version of the link between rainbows and serpents is found throughout Australia. The most important deity of Australian Aborigines is the Rainbow Serpent. According to some accounts, the Rainbow Serpent came down from the sky during the Dream Time, a period at the beginning of creation when huge spirit beings journeyed across the land and filled it with plants, animals and the souls of everyone who will ever be born. The Rainbow Serpent is said to live in deep water holes. During the dry season it sucks up water from the earth that it later expels as rain during the wet season. When visible as a rainbow, the Rainbow Serpent is said to be travelling between waterholes. Although it has almost as many names as there are tribes, for all Aborigines the Rainbow Serpent is not so much a creature as the embodiment of the creative and destructive powers of nature, particularly those due to rain and water.

The widespread identification of rainbows with serpents is almost certainly due to the prevalence of large, highly coloured snakes—pythons in equatorial Africa and anacondas in Amazonian South America—that inhabit water holes and swamps in those jungles. Their brilliant colours, aquatic habits, ability to climb trees and, perhaps most importantly, the fact that these snakes reappear after hibernation at the start of the rainy season, may all have suggested and reinforced a link between serpents and rainbows.[49]

The wide distribution of the rainbow serpent motif is almost certainly the result of human migration rather than of multiple spontaneous inventions. The belief that the rainbow is a giant serpent is thought to be indigenous to Equatorial Africa, from where the idea was taken to other parts of the world by humans as they spread across the globe. During the last ice age, when sea-levels were perhaps 100 m lower than they are at present, land bridges joined Sumatra and Borneo to mainland South east Asia, making migration possible across lands that have since been separated by sea.

And early in that migration the rainbow snake took on another form: that of a dragon. The idea that dragons are the arbitrary, independent creation of

---

[47] Lévi-Strauss, C. (1970), pp 246–7.
[48] Urton, G. (1981), p 115.
[49] Urton, G. (1981), p 116.

overactive imaginations of people widely separated in time and space has been contested by Robert Blust (b.1940), a professor of linguistics. He argues that one has only to draw up a list of the salient features of dragons to see the link, for wherever these mythical creatures are found, be it in Europe or in Asia, they are invariably associated with water in the form of pools, springs, rivers and waterfalls, a characteristic shared with the Rainbow Serpent. Dragons also control the weather, and are responsible for both rain and drought.

> By far the most common view is that the rainbow is a giant snake which either drinks water from the earth and sprays it over the sky (thus causing it to rain), or that drinks rain from the sky (thus causing it to stop).[50]

That is not to say that dragons in China are identical in every respect to those in Europe. While all dragons are associated with water in one form or another, they vary in other ways. European dragons are famously demonic and must be overcome by a hero, whereas in China a dragon denotes vigour, optimism and prosperity. The best-known dragon slayer is, of course, St George and the forerunner of his encounter with the dragon and its outcome is almost certainly the story of Marduk and Tiamat.[51]

Another telling similarity between rainbows and dragons is their association with hidden treasure. The end of the rainbow has throughout history been invested with particular meaning and has played an important role in the history of religions and civilizations. "A common belief in European folklore is that a monstrous creature or reptile, like a dragon, is guarding hidden treasure at the rainbow's end."[52] Blust supplies a plausible explanation for this belief.

> Why do dragons guard treasures? Because there is gold at the end of the rainbow. This answer may seem provocative or facetious, but the connections are reasonably straightforward. The end of the rainbow is widely believed to touch a water source, typically a spring or river. Unlike most metals, gold is commonly found in small amounts in alluvial river washings, where it can be easily seen with the naked eye. For this reason there can be little doubt that gold was the first precious metal known to early man long before the advent of metallurgy, a fact that may partially account for its peculiar salience in myth and psychology. Since the rainbow touches down in a spring or river, its ends covers a place where gold is found. And since the rainbow is a giant serpent

[50] Blust, R. (2000), p 525.
[51] Qiguang Zhao (1992), pp 36–9.
[52] Oestigaard, T. (2019), p 47.

which guards springs and rivers when it does not appear in the sky, it guards the gold found there.[53]

All that remains to us of the age of myth, animated as it was by a pre-scientific belief that nature is imbued with the divine, are quaint stories preserved in ancient myths or unearthed by anthropologists studying isolated communities. We may find these stories charming, amusing, even instructive, but we can never participate in them wholeheartedly. We lack that spontaneous, unconditional, pre-scientific engagement with nature that gave mythologies such sway over the imaginations of our ancient fore-bearers. Nowadays people consider the rainbow a rather jolly, optimistic sight, an occasion that brings a smile to the lips, a momentary "…my heart leaps up when I behold/a rainbow in the sky." But myths about rainbows show us that it wasn't always so. More often than not the rainbow was a baleful portent that induced anxiety and dread in those who saw it.

Perhaps as a consequence of its many baleful associations, folklore concerning the rainbow is full of advice on how to avoid its malevolent influence. Within living memory, Shetland Islanders regarded a rainbow over a house as presaging the imminent death of one of its occupants, clearly a reference to the rainbow's role as a soul bridge. And it was once widely considered reckless to point at a rainbow with one's index finger because doing so would result in the offending finger withering or falling off; it might even lead to the death of a close relative. Should one wish to draw attention to a rainbow without arousing its displeasure, one was advised to do so by gesturing with a nod of the head, a closed fist or an elbow. The prohibition has been called the rainbow taboo.[54]

According to Bernabé Cobo, the Incas "did not dare to look at [the rainbow], or if they looked at [it], they did not dare to point a finger at [it], believing that they would die; and to that part where it seemed to them that the foot of the arch fell, they considered it a horrendous and fearful place, understanding that there was some guaca or other thing worthy of fear and reverence."[55] A guaca is a burial site or tomb.

Why anyone should avoid pointing at a rainbow may have its roots in the universal human aversion to being pointed at by another person. Pointing with an outstretched index finger is a peculiar to humans, an action that is invaluable in drawing the attention of others to this or that feature in one's surroundings. But pointing can also be a hostile act, as when accusing or

[53] Blust, R. (2000), p 532.
[54] Blust, R. (2021), p 145.
[55] Cobo y Peralta, B. (1893), p 149.

belittling someone. Pointing at the malevolent spirit embodied by a rainbow is surely asking for trouble. Atonement involves wetting the offending finger by inserting it in one's mouth or anus.

One of the few remaining examples of practical folklore about the rainbow has already been mentioned: "Rainbow in the morning, shepherd's warning, rainbow in the evening, shepherd's delight".

There is a nautical variant of this:

Rainbow in the morning, sailors take warning;
Rainbow at night, sailors' delight;
Rainbow to windward, foul fall the day;
Rainbow to leeward, damp runs away.

Modern meteorology broadly supports both these pieces of weather folklore.

But there are several examples of folklore about the rainbow supposedly rooted in experience that are less convincing, such as the belief that its colours are signs of the potential bounty of a harvest. For example, if the red is dominant in a rainbow then the vineyards will produce a good crop and wine will be plentiful. If the dominant colour is yellow the grain harvest will be bountiful. If green is particularly marked, it signifies that there will be enough hay for livestock during the whole year. But the colours of a rainbow depend on local conditions that usually last for a few minutes and so can't be a sign of events that lie weeks or months, let alone a single day, in advance.

The ancient Chinese classified rainbows according to colour and shape and saw these as omens of bad luck. This made it possible to employ a rainbow as an augury for such things as political fortune, marital infidelity or the success or failure of crops. But as far as its nature was concerned, they saw it as combination of yin and yang, the universal male and female principles. The clearest expression of these ideas was given by the Chinese scholar, Hsing Ping (932–1010 A.D.):

The rainbow appears more frequently in spring. When a pair of rainbows appear simultaneously, the brighter one is the male rainbow, the fainter one is female. The male rainbow is the combination of yin and yang. If there is only yin or yang there will be no male rainbow. When the cloud is thin and when the solar rays reflect from the raindrops, then the male rainbow appears … the female rainbow has a curved shape. The male rainbow is green and red, or white; it is yin.[56]

---

[56] Sayili, A. M. (1939), p 83.

Folklore can sometimes seem to be a half-way house between myth and science. Compared to myth, folklore can seem to be based on a pragmatic, common sense view of the world. But it would be a mistake to assume that it is a precursor of science. For a start, it is insufficiently systematic and lacks a reliable causal scheme. Science requires more than observation, it also calls for disciplined, rational speculation.

Fittingly, the transition from myth to science took place among the peoples for whom Iris was the very embodiment of wonder.

4

From Myth to Mathematics

I see, my dear Theaetetus, that Theodorus had a true insight into your nature when he said that you were a philosopher, for wonder is the feeling of a philosopher, and philosophy begins in wonder. He was not a bad genealogist who said that Iris is the child of Thaumas.[1]

Plato's claim that making Iris the daughter of Thaumas is an acknowledgement that the rainbow is the embodiment of wonder has stood the test of time. Thaumas, in case you don't know, is also the Greek word for wonder. As for his daughter, well, as we have seen in the previous chapter, she was considered the personification of the rainbow by Greeks and Romans alike.

These days, regrettably, wonder is all too often held to be the preserve of the heart rather than the head. And in some quarters, explanation, particularly scientific explanation, has come to be regarded as anathema to wonder. "Don't spoil it with explanations," pleads the romantic, "just enjoy it." But genuine wonder is not the state of dumbstruck, paralysing amazement of a credulous and unlettered clod faced with a hitherto unknown spectacle that this sentiment implies. It is, as Plato knew, the midwife to science, the necessary prelude to discovery and investigation and a spur to inquiry and speculation.

The first person to speculate about the nature of things in a manner that went beyond an appeal to supernatural forces was a rather shadowy figure called Thales. According to the few fragmentary accounts that we have of

---

[1] Plato (1952), 155d, p 55.

© The Author(s), under exclusive license to Springer Nature
Switzerland AG 2023
J. Naylor, *The Riddle of the Rainbow*, Copernicus Books,
https://doi.org/10.1007/978-3-031-23908-3_4

him, he was born around 600 BC in Miletus, the major city of Ionia, a Greek colony on the Aegean coast of modern Turkey, and seems to have been a merchant who acquired a knowledge of mathematics and astronomy during his travels. He is said to have predicted an eclipse in 585 B.C.—extremely unlikely given how little was understood about the relative motions of the earth and moon at the time—and to have introduced his fellow Greeks to geometry—a plausible claim given that he may well have spent time in Egypt. But his fame rests principally on the claim that he is supposed to have asked questions that myths don't or can't answer, questions like "of what are things made?".

His answer was that the elemental substance is water. Unfortunately, only snippets of his thoughts have survived, so we can't be sure how or why he came to this conclusion.[2] Scholars now suggest that what made water a plausible candidate as nature's universal building block for Thales is that as well as being essential to all living things, it is also the only substance that can exist on earth as a solid (ice), a liquid (water) and a gas (water vapour). But it is also possible that he may have been influenced by a common theme of the myths about the creation of the world. As we saw in the last chapter, according to the *Enuma Elish*, Marduk divided Tiamat's watery body to create the heavens and the earth, a story that was probably widely known throughout the Middle East in Thales day. Nor was the idea that the world was originally created from water confined to the Babylonians. In Genesis we learn that on the second day of creation God separated the waters that were under the firmament from those above it and that on the third day he separated the waters from the land. Moreover, primordial gods of Mesopotamia, Egypt and Greece: Ea (Mesopotamian), Nun (Egyptian) and Okeanos (Greek) are all associated with water.

Some fifty years later, another Milesian, Anaximenes (586–526 BC), claimed that water is much too tangible to be the primary substance; only air can fill this role. In the first place, he said, air is essential to life because no creature can survive without breathing. As for the material world, he went on, water is really condensed air, which, when further condensed, first becomes earth and ultimately rock. He even devised a mechanism to explain how air could change from a gas to a solid and back again. In what must count as the earliest example of the use of an experiment to illustrate a physical principle, he came up the following demonstration. With your mouth wide open, breathe on the back of your hand: your breath feels warm. Repeat the experiment, this time blowing hard though pursed lips and your breath feels cooler.

---

[2] For the little we know about Thales' views on the nature of matter see Aristotle (1933), Book 1.

According to Anaximenes, this shows that heat is really rarefied air and that cooling air brings about condensation.

These Milesian thinkers, called *physiologoi* by a later generation of Greek thinkers, we now call *natural philosophers* because they speculated about the nature of things without invoking supernatural causes.[3] But we should not think of them as diehard materialists who completely denied the gods any role in the scheme of things. Indeed, Thales is supposed to have asserted that the gods are present in all things. So, while not entirely abandoning supernatural interpretations of natural events, the observation and speculation that the search for the essence of matter demanded meant that the *physiologoi* focused their attention on the material world, thereby laying the foundations of Greek science of later centuries, which culminated in the uncompromising materialistic atomism of Democritus (460–370 BC).

Democritus maintained that the universe consisted entirely of "atoms and the void." Atoms were conceived as tiny, unbreakable particles in constant motion within a void (i.e. empty space or vacuum). Unfortunately for the development of science, these ideas were not followed up by later generations of Greek thinkers, and atomism, the most promising basis for a systematic account of the properties of matter that the ancient world came up with, was more or less entirely forgotten until middle of the seventeenth century. Indeed, without the revival of atomism during that century, it is difficult to see how science as we now know it could have come about. We'll revisit this claim in later chapters.[4]

Anaximenes was also the first natural philosopher to pronounce on the rainbow: he said that it is a reflection of the sun by a cloud. As explanations go, this doesn't get us very far, though it's a start. If nothing else, it broke with the age-old belief that the rainbow is either a sign of supernatural forces at work or a manifestation of a mythical being. But as with all things, one question leads to another. How can something as insubstantial as a cloud reflect sunlight in such a way that it creates a multicoloured arc? Unfortunately, Anaximenes left no written works, and as far as we know the only other thing he said on the subject was that the colours in the rainbow are due to a mixture of sunlight and darkness, an association that was to bedevil thinking about colour until the turn of the nineteenth century.

And there the matter rested for the next two centuries until it was taken up by one of the cleverest and most influential thinkers the world has known, the philosopher Aristotle. Aristotle was a philosopher at a time when philosophers

---

[3] *Physiologoi* can be translated as *physicists* in English.
[4] The natural philosophers of the seventeenth century would have learned of Democritus' ideas from the first two parts of "The Nature of Things" by the Roman poet Lucretius.

speculated about much more than the meaning of life or the nature of truth. But even among his peers, Aristotle's intelligence, curiosity, and knowledge were legendary. His scientific interests alone spanned biology, cosmology, meteorology and physics. Nor did his reputation rest on having made himself master of all that was known in his day, he was also the first thinker to identify and systematically develop entirely new branches of knowledge such as metaphysics, logic, ethics and politics.

Aristotle was born in 384 BC in Stageira, a city in Macedonia. His father, Nicomachos, was physician to Philip II, the King of Macedon and father of Alexander the Great. When Aristotle turned seventeen he was sent to Athens to complete his education. He spent the next twenty years there, much of it as a student of Plato, arguably the greatest philosopher of ancient Greece. When Plato died in 348 BC, Aristotle left Athens and spent the next few years on the island of Lesbos developing his own ideas and carrying out zoological research. He was recalled to Macedonia in 343 BC to tutor the young Alexander. Aristotle returned to Athens in 335 BC where established a philosophical movement based on his ideas and which came to be known as the Peripatetic School because, it is supposed, Aristotle delivered his lectures while strolling through the colonnades of the building where he taught his students.[5] But, following Alexander's death, he had to leave the city in 323 BC when Alexander's enemies turned on anyone who had enjoyed his patronage. Aristotle died the following year, aged 62.

Most of Aristotle's writings have been lost. What survives consists largely of notes of uneven quality that were gathered together, edited and eventually published long after his death. They are clearly not finished works intended for publication. If anything, they are works in progress and don't necessarily represent his final thoughts. In some cases they may even be notes taken down by his pupils when he was teaching them, something to bear in mind when wrestling with his account of the rainbow, which is to be found in his work on meteorology.

For the Greeks of Aristotle's day, meteorology was the study of meteors, or 'things on high', from *meteoros*, which is Greek for 'to be raised up'. Meteors were considered to be events that occur within the atmosphere, which according to Aristotle extended from the earth's surface to the orbit of the moon. When Aristotle got around to meteorology, in characteristic fashion he set about collecting, classifying and explaining every type of meteor known at the time. The fruits of his labour are gathered together in a work entitled *Meteorologica*, composed sometime around 350 BC.

---

[5] The building was known as the Lyceum.

Although Aristotle's meteorology dealt with many of the events that we associate with weather today, such as rain and snow, thunder and lightning, and clouds and winds, it included others that we no longer recognise as being weather-related, such as earthquakes and the saltiness of the sea. He also classed comets and shooting stars as atmospheric events rather than astronomical ones.

To understand why he did so it helps to know something of his cosmology. Aristotle rejected the atomism of Democritus, with disastrous consequences for the subsequent development of the sciences. One of the reasons he did so was that he believed that the void, the completely empty space necessary if atoms are move about freely, is an absurdity. His main objection was based on his assumption that the natural state of a body is rest and that without a constantly applied force to drive it forward a body will either remain stationary or quickly come to a halt. Indeed, there are any number of everyday experiences that appear to confirm this claim: a force must be applied and maintained to lift an object, stop peddling and your bicycle slows down, stop paddling and your canoe drifts to a stop. But in a vacuum there is nothing to oppose the force necessary to initiate or maintain motion. Aristotle concluded that in a void the slightest push would cause a body to move with infinite speed: a stone would fall to the ground in an instant, whatever height from which it was dropped.

He also disputed the atomists' claim that the universe is infinite in extent. In place of the Atomists' infinite void in which things are made of tiny indivisible particles, Aristotle proposed that the universe consists of two distinct realms, one within the other. The inner realm contains a stationary, spherical Earth together with its atmosphere, and is known as the sublunary sphere because it lies within the moon's orbit. Beyond this, in all directions, lies the realm of stars and planets, which are composed entirely of *quintescence*, an element not present in the sublunary sphere. The sublunary sphere itself is composed of a combination of four elements: earth, water, air and fire. When mixed in different proportions, these not only account for the variety of things around us, but also explain why the sublunary sphere is subject to change and decay. The starry realm, on the other hand, is incorruptible and eternal: it can't undergo change because it is composed of a single element.

Given that he believed that the starry realm is fixed and unchangeable, Aristotle erroneously concluded comets and meteors must be events within the atmosphere because they are short-lived and so involve change. Ironically, the word meteor, the generic term that Aristotle applied to all atmospheric events, and which was used in that sense up until the twentieth century, is today usually used only in connection with an event that has an extra-terrestrial origin: a meteor (or shooting star) is the luminous trail seen when a small fragment of interplanetary debris heats up as it plunges through the atmosphere.[6]

One of the meteors that Aristotle studied in great detail was the rainbow. As was his practice, he began with a detailed description of the phenomenon. He pointed out that the arc of a rainbow is a semicircle and that how much the arc is visible depends on the height of the sun above the horizon. It's a full semicircle when the sun is at the horizon, but as the sun rises, the amount of the arc that is visible shrinks until it disappears altogether when the sun has attained a certain height. He didn't say how high the sun must be before a rainbow is no longer visible but he did point out that the reason why we don't see rainbows in the middle of the day in summer is that the midday sun is too high in the sky to form a bow.[7]

So far, so good. But at this point Aristotle muddied the waters by adding that the radius of a rainbow depends on the height of the sun above the horizon. He claimed that the radius is least when rainbow is a full semicircle, i.e. at sunrise or sunset. In fact, as we now know, the radius of the rainbow arc can vary from one occasion to another, though this has nothing to do with the height of the sun. The circumstances that bring about changes in the rainbow's radius—it's to do with the size of raindrops—were not fully understood until well into the nineteenth century, 2300 years after Aristotle. He was probably misled by the moon illusion. As we noted in chapter two, just as the moon appears to be larger than usual when it is seen close to the horizon, a rainbow appears to have a larger radius when it's arc hugs the ground, which happens when the sun is well above the horizon. Aristotle was aware of the moon illusion, and refers to it in his account of the rainbow in connection with the sun's apparent enlargement when it's on the horizon. But it didn't occur to him that this might also explain the apparent change in the rainbow's radius.

He went on to say that there are two bows, the inner one being brighter than the outer one, and that although the colours in both bows are the same, red, green and blue, the order of colours in the outer bow is the reverse of that in the inner bow. This is the first time that the secondary bow had been written about, let alone described.[8] In another first, Aristotle noted that rainbows are also formed by moonlight. He claimed that the resulting moon-bows are rare because they can occur only at full moon, i.e. once a month,

---

[6] Proof of the extraterrestrial origin of meteors was finally established in 1794 by Ernst Chladni (1756–1827).

[7] Aristotle (1952), chap. II, p 241.

[8] Aristotle (1952), chap. II, p 243.

Fig. 4.1   A moonbow over the town of Kihei, seen from Kula, on Maui, Hawaii, US. The reason you can see colours in the bow is because the photograph is a time exposure[10]

when moonlight is bright enough to produce a visible bow. Furthermore, he believed that a moonbow would form only when the moon is close to the horizon, either rising or setting. All things considered, he said, it's not surprising that there are so few reported sightings of moonbows; indeed, he was aware of only two such reports in fifty years (Fig. 4.1).[9]

Considering that this is the first time the rainbow had been described in any detail, what Aristotle has to say about it can scarcely be faulted. Alas, despite such a promising beginning, the explanation that followed failed to account correctly for a single one of the rainbow's features. It could hardly have been otherwise because Aristotle's ideas about vision, light and colour are all completely mistaken.

Aristotle accepted that vision involves light. But Aristotelian light doesn't pass from the object to the eye. Instead, he says, light enables us to see the shapes and colours of objects because it makes air transparent. For Aristotle, light is a state of being, and not as it is for us, a palpable form of energy that activates vision.

[9] Aristotle (1952), chap. II, p 245.

[10] Photo by Arne-kaiser, Wikimedia Commons, https://tinyurl.com/5n726eyj Accessed 1/10/22.

Unaware that light and colour are inseparable, something that was not firmly established for another two thousand years, he took what to most people is the common sense view: that colour is an intrinsic property of an object and that light merely makes colour visible without altering it in any way. Hence objects retain their colours even in complete darkness, something, indeed, that most of us still take for granted despite the fact that it has no scientific basis. As we shall see in the next two chapters, the weakness of Aristotle's ideas about colour wasn't exposed until Isaac Newton deployed his prims in the seventeenth century.

But what of the colours of a rainbow? Aristotle realised that a rainbow is the reflection of the sun, not an object in its own right, which led him to conclude that its colours are not real, unlike those of objects. Instead, he attributed them to changes in brightness. He had noticed that on some occasions that "when we look directly at a cloud that is close to the sun, it appears to have no colour but to be white, but when we look at its reflection in water it seems to be partially rainbow coloured."[11]

Thin clouds that are close to the sun are indeed often markedly iridescent. Aristotle attributed the iridescence to a reduction of a cloud's brightness that occurs when it is reflected. He didn't realise that the iridescence is produced within a cloud and that the reason we can't easily see the colours directly when looking at the cloud is that the eye is dazzled by the sun. In fact, the colours become visible because reflection reduces the brightness of both the sun and the cloud.[12]

In support of the idea that in some situations colour depends on brightness he pointed out that no colours are brighter than white or darker than black. Moreover, he insisted, when colours are compared to one another we find that red is brighter than green, which in turn is brighter than blue. It follows, he argued, that colours can be generated by altering the brightness of white light, or, to use his terminology, colour arises when 'vision is weakened'. Red, he said, is brought about when this weakening effect is least, blue when it is greatest.

Aristotle's ideas about the colours in a rainbow are superficially plausible. Making a colour darker does indeed alter its appearance, but what has changed is its brightness not its hue, though it sometimes seems as if a new colour has been produced. Shades of brown, for example, can be created from red or orange by reducing their brightness; but it's not possible to create green

---

[11] Aristotle (1952), chap. IV, p 261.
[12] These iridescent colours are due to diffraction of sunlight by cloud droplets.

or blue by making red darker.[13] However sympathetically one interprets Aristotle's theory of colour, it simply doesn't account for the range of colours seen in a rainbow.

Aristotle's chances of successfully explaining the rainbow were further reduced by a widespread disdain for experiment among Greek thinkers. By and large, the natural philosophers of the time seldom performed experiments either to test their ideas or to clarify them and all too often based their explanations on untested speculation. The widespread prejudice against experiment sprang, in part, from a belief that it is better to wait passively and see what Nature reveals than to force the issue artificially with an experiment. A student of nature, asserted Aristotle, must be a patient and observant onlooker, not a meddlesome experimenter. There was also an element of prejudice: only slaves and craftsmen use their hands whereas intellectuals use their heads.[14]

Passive observation often pays dividends in biology and zoology, fields in which Aristotle excelled and whose discoveries about marine life were not, in some instances, equalled until well into the nineteenth century. But where the rainbow is concerned, observation alone proved insufficient; what was needed was an experiment or two with a transparent crystal sphere. Aristotle would have known that transparent crystals sparkle with the same colours as are seen in a rainbow, yet he never made the connection. Thus, when it came to explaining the rainbow, in the absence of insights that he might have gained from such experiments, he relied entirely on geometry and assumed that the key to the rainbow lies in a particular geometrical relationship between the sun, the rainbow arc and the observer. In fact, there is such a relationship, but it's determined by the passage of light through drops of water, which is why any attempt to explain the rainbow has to begin with experiments either with a crystal sphere or a spherical flask filled with water. By starting out from the wrong place, Aristotle's explanation fell at the first hurdle.

At this point you may be wondering whether it is worth struggling through the rest of Aristotle's account if it leaves us no wiser about the nature or cause of the rainbow. Since he failed to provide a viable explanation for any of the rainbow's features, why not fast forward to thinkers of later generations who came closer to the truth? The answer is that Aristotle had an enormous influence on almost everyone who wrote on the subject until the close of

---

[13] See this for yourself on a computer screen by drawing a shape in a painting application and altering its colour using the RGB sliders of the colour pallet.

[14] This prejudice survived into our own era. Many of the leading natural philosophers of the Scientific Revolution such as Descartes and Huygens looked down on the artisans who made the instruments on which they relied for their researches and were reluctant to share their ideas about the theoretical basis of their instruments with them.

the seventeenth century, almost two thousand years later. Another reason is that he was far more thorough than most of his successors. Not only was his description of the rainbow's features almost spot-on, he also attempted to account for these features, a real tour de force, something that few others managed to achieve before the seventeenth century.

Aristotle agreed with Anaximenes that a rainbow is "a reflection of our sight to the sun" by a cloud, something that was generally accepted among Greek thinkers of the time. According to the prevailing view, however, the shape of the rainbow's arc was attributed either to the roundness of the reflecting cloud, or to the fact that the rainbow is a reflection of the sun, which is itself a circle. Aristotle rightly rejected both these views and set out to prove that the shape of a bow is determined by the relative position of the sun, the observer and the cloud in which the bow is seen.

Already there are grounds for confusion. How can sight be reflected to the sun? Surely, it's the other way around? Light reaches the eye from the sun. In fact, the belief that vision depends on the eye emitting a visual ray was widely held among Greek natural philosophers—it's known as *extramission*. The opposite idea, that light enters the eye from outside, is known as *intromission*. Elsewhere in Aristotle's writings we find that he favoured intromission as the basis of vision, so his use of extramission in his explanation of the rainbow is doubly problematic. Extramission has an undeniable emotional appeal; we still speak of casting an eye over a scene, of a penetrating gaze and of a look that could kill. But, physically, extramission is untenable, impossible to defend rationally, and an idea that Aristotle firmly rejected in most of his writings on optics. Perhaps we have here one of the inconsistencies that lend support to the possibility that it was an inattentive student scribe rather than Aristotle himself who is responsible for the text of *Meteorologica*. As we shall see, however, the real weakness of Aristotle's explanation has little to do with whether vision is a matter of intromission or extramission.

One of the most compelling illusions that confronts us when we are outdoors during daylight is that the sky appears to be a vast flattened dome that arches above us and which meets the earth at the horizon. Aristotle made this illusory dome central to his explanation of the rainbow. He set the scene by imagining an observer at the centre of a large, level plain over which arches the hemispherical vault of the sky. This sky-vault played the same role in his explanation of the rainbow as the celestial sphere does in astronomy. It's a simplifying device that allowed him to single out what he considered to be the most important features of the situation: the relative position of the sun, observer and the cloud in which the rainbow was seen. He placed the sun on the horizon and the cloud on the opposite side of the sky, choices that made

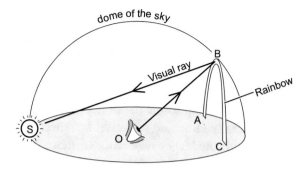

Fig. 4.2 Aristotle explanation for the shape of a rainbow. The visual ray from the eye at O to the sun S on the horizon traces out a semicircle ABC on the dome of the sky

it impossible to arrive at the correct explanation because it places the observer midway between the sun and the bow and so doesn't represent accurately the geometry of the actual situation (Fig. 4.2).

Nevertheless, Aristotle correctly pointed out that if the angle between the visual ray that joins the eye to rainbow and the one that joins the rainbow to the sun is constant, then, as a matter of geometrical symmetry, the rainbow must be a semicircle because under these conditions the point on the vault of the sky at which the reflection of the sun occurs traces out a circular arc on the imaginary sky dome. Unless you are good at visualising three-dimensional spaces it is difficult to understand this point without a diagram.

Establishing a plausible explanation for the rainbow's semicircular arc on the basis that it is a reflection from the sky dome proved to be the high point of Aristotle's explanation. It was downhill from there as he grappled unsuccessfully with questions that weren't properly answered for another two thousand years. What determines the size of the arc? Why does the reflection of the sun responsible for the rainbow occur where it does? Why is the reflection confined to a narrow band? And the question that was to prove the most challenging until Newton found the answer sometime in 1660's: what is the origin of the rainbow's colours?

Answers to these questions go to the heart of the matter of the how a rainbow is formed. This, as we now know, is due to what happens when light passes through a drop of water, something that can't be understood without detailed understanding of reflection and refraction.

Reflection is the change in direction that occurs when light meets a surface and is redirected away from it, like a ball bouncing on a hard surface. Refraction is the change in direction that occurs when light passes from one medium to another, say from air to water or vice versa. These are matters of everyday observation that would not have escaped someone as observant as Aristotle. Unfortunately, he often used the terms interchangeably, which suggests that he didn't have a clear idea of the difference between them. In any case, by locating the rainbow on the surface of the inverted, hemispherical skydome that is centred on the observer, the visual ray that Aristotle says is emitted by the observer's eye towards the rainbow travels along a radius to the circumference of the skydome and so would be reflected directly back to his eye in accordance with the law of reflection. In short, it is impossible for Aristotle's visual ray to reach the sun.

Not that he invoked the law of reflection to find the position of the rainbow on the surface to the sky dome. Instead, he used a geometrical proof based on proportions. He said that the reason the primary rainbow occurs where it does is due to the ratio of the distance from the sun to the rainbow to that of the distance of the rainbow to the observer. He suggested that this has a particular value but neglected to say what it is, though he could have done so, given that the angular size of the rainbow can be directly measured with simple apparatus. Aristotle's proof is not difficult to follow if your knowledge of geometry is up to scratch, but, since it's rather involved, and because the location of the rainbow relative to the observer can't be determined in this way, we can safely ignore it.

It may be that Aristotle didn't make use of the law of reflection because he believed that that there are two kinds of reflections. He pointed out that large, smooth surfaces are able to reflect the shapes and colours of object, which is why we see their images when we look into a mirror or a pool of water. But, according to him, there is another type of reflection, one in which colour alone is reflected. Perhaps he had in mind what we see when sunlight shines on a dewdrop: it sparkles with colours rather than reflecting a single bright, uncoloured image of the sun. This is why we can see the rainbow's many colours only in a multitude of raindrops. Or, to put it in the sort of language he might have used: a cloud of raindrops reflects only the sun's colours and not the sun itself. And, said Aristotle, this multi-coloured reflection is the rainbow.

But when it came to the order of colours, Aristotle, the master logician, made a surprisingly elementary error in logic and begged the question because he explained the sequence of colours in the primary bow in terms of the size of each of the coloured bands. The outer band being the largest, he said, is brighter than the others and must therefore be red. The innermost band, being the smallest, is darker than the others and so must be blue. Using the same reasoning, the intermediate band is green. Yellow, which Aristotle admitted is seen between the red and green arcs, he attributed to an illusion due the proximity of red and green (Fig. 4.3).

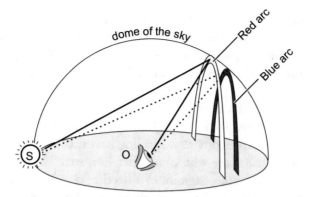

Fig. 4.3 Aristotle's explanation for the colours seen in the primary rainbow based on his account in *Meteorologica*. The primary bow is formed on the surface of the dome of the sky

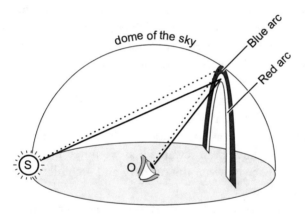

Fig. 4.4 Aristotle's explanation for the colours seen in the secondary rainbow based on his account in *Meteorologica*. The secondary bow is formed on a vertical plane, not on the dome of the sky

But the same explanation can't be used for the sequence of colours in the secondary bow because they are the other way around to those in the primary bow. Instead, he said, these are determined by the distance travelled by the visual ray between the eye and the sun. Because the inner band of the secondary is red, it is due to a reflection of the sun that has not been weakened to the same degree as the reflection responsible for the outer blue band. This, in turn, requires that primary and secondary bows are formed at different distances from the observer (Fig. 4.4).

Even though Aristotle's explanation of the rainbow proved to be wrong in almost every respect, its importance is that it was the most thorough account of the phenomenon known to the ancient world. It gave heart to those timid souls for whom the phenomenon might otherwise have seemed dauntingly beyond human reason and became, if not the template, then a point of departure for almost all explanations of the rainbow for the next two thousand years.

Unfortunately, uncritical acceptance of Aristotelian ideas meant that his account of the rainbow, particularly given the Greek aversion of experiment, acted as a brake to further progress. Reading later accounts of the rainbow, it becomes obvious that those who came after him were unable to free themselves from an Aristotelian quagmire of reflecting clouds, of colours brought about by the darkening of the visual ray emitted by the eye, or the fiction that the rainbow is formed on the surface of the vault of the sky.

Mesmerised by Aristotle's intellect, the best they could achieve was to follow his explanation to its logical conclusion. And in doing so, some five hundred years after Aristotle's death, Alexander of Aphrodisias (f.200AD), a Peripatic philosopher, noticed a logical inconsistency in Aristotle's explanation of the rainbow. If red is the brightest colour, Alexander reasoned, then the gap between the outer red segment of the primary bow and the inner red segment of the secondary bow should not only be red, it should also be the brightest part of sky when there is a double rainbow. Yet the gap is both colourless and noticeably dark. Of course, committed as he was to Aristotelian methods and ideas, Alexander could offer no explanation for this fact. But his efforts have been rewarded: as we noted in chapter, the absence of colour in the space between the primary and secondary bows is known as Alexander's dark band.[15]

It would have been far better if he and every other scholar had torn up their copy of *Meterologica* and tackled the problem afresh. In fact, as we shall see, as the thirteenth century gave way the fourteenth century two natural philosophers did just that, and were amply rewarded by discovering that the secret of the rainbow lies in what happens to light inside a drop of water.

Greek philosophy and science ran out of steam long before the Roman Empire began to break up during the fifth century as a result of repeated barbarian onslaughts. The consequences for intellectual life of this fragmentation was made worse by the fact that the language of science and philosophy in the Roman world was predominantly Greek, and that the centre of gravity of learning lay in the cities of the Eastern empire such as Athens and Alexandria. The decline and eventual collapse of the Western Roman Empire

---

[15] The darkness of the *dark band* is due to the absence of light reflected from drops reaching the eye of the observer. It looks dark in comparison to the brightness of the two bows. Descartes gave the correct explanation for this in 1637.

isolated Europe from the bulk of Greek learning, including most of Aristotle's works, because few of them had been translated into Latin. Scholars in Western Europe remained ignorant of almost all Greek science, mathematics and philosophy from the seventh to the eleventh centuries, a period known in the history of Europe as the Dark Ages. During these centuries, the Church held a monopoly of learning and literacy, with the consequence that scholarship became a handmaiden of theology. And without access to Aristotle's *Meteologica*, speculation about the rainbow all but ceased.

The Eastern Roman Empire also came under attack. Following the death of the Prophet Mohammed in 632 AD, Arab armies overran and conquered most of the lands of the Eastern Roman Empire where the tradition of Greek learning and culture had continued to flourish. When the dust of battle finally settled in the eighth century, the Arabs found themselves heirs to Greek science, mathematics and philosophy. They commissioned translations of Greek texts from their Christian subjects, enabling Arabian and Persian scholars to study, expand and build upon the achievements of earlier philosophers, astronomers, mathematicians and doctors.

Given the situation, it's not surprising that the most original natural philosophers towards the end of the first millennium were to be found in Muslim-dominated lands. And the greatest of them was an Arab, Ibn al-Haytham (c.965–c.1040), known in Europe as Alhazen. Alhazen was born in Basra but spent his adult life in Cairo, where he laid the foundations of modern optics and wrote his *Book of Optics*.[16] All but ignored in the Islamic world, this work became the most influential book on the subject in Europe until the end of the sixteenth century. Among other things, Alhazen was a pioneer of the experimental method of discovery, performing experiments to test and develop his ideas. Among his many achievements he proved that light is emitted by luminous sources such as flames and the sun, and is reflected from objects into the eye, thereby demolishing the idea of extramission. and its associated visual ray. But, disappointingly, he had little to say that was novel on the subject of the rainbow. Despite experimenting with water-filled glass spheres, he accepted Aristotle's ideas that a rainbow is due to reflection alone and that its colours are due to the darkening of light.

Another major Arab thinker, Ibn Sina or Avicenna (980–1037), who earned the title of the 'Arabic Aristotle' because of his wide learning, disagreed with Aristotle about the location of the rainbow. He had seen a rainbow against a hill and concluded that a bow was not a reflection from a cloud

---

[16] Ibn Al-Haytham, 1989.

but from a transparent medium lying between an observer and the cloud. He was also critical of Aristotle's account of colour but offered none of his own.

Northern Europe remained an intellectual backwater until the twelfth century when European scholars first began to get their hands on the works of Greek, Arab and Persian natural philosophers and mathematicians, principally through translations into Latin of Arabic versions of the original Greek texts. Among these was Aristotle's *Meteorologica*, unknown in Western Europe until it was translated into Latin around the middle of the twelfth century by the most prolific translator of his age, Gerard of Cremona (c.1114–1187). With the Latin edition of *Meteorologica,* interest in the rainbow among European thinkers received a fresh impetus. The other important work to be translated at this time was Alhazen's *Book of Optics*, which quickly became far better known and studied in Europe than it ever was in Islamic lands.

But hardly had the ink dried on the parchments of the first translations of these works than dissenting voices were heard. The first European thinker to disagree with Aristotle about the cause of the rainbow was Robert Grosseteste (c. 1175–1253), an English theologian at Oxford University. He pointed out that sunlight reflected from a cloud results in the whole cloud being bright, not in a concentration of light in a narrow arc or band. In any case, if a rainbow were a reflection from a concave cloud, as Aristotle seemed to imply, the reflection would move up and down with the sun, whereas observation confirms that it rises as the sun drops towards the horizon and vice versa. Grosseteste's own view was that a rainbow is due to refraction in the medium between the sun and the cloud in which the rainbow is seen. A cloud merely acts as a screen on which the rainbow is projected. Shortly after publishing his account of the rainbow in 1235, Grosseteste was appointed Bishop of Lincoln. As bishop he opposed the Pope's attempt to foist Italians on the English religious establishment, but was eventually excommunicated for his pains.

Roger Bacon, (1214–1292), a Franciscan friar, who like Grossesteste also fell foul of his ecclesiastical superiors, was a renowned pioneer of the new natural philosophy and had a particular interest in optics. Bacon rejected Grosseteste's account of the rainbow. But his lasting contribution to our understanding of the rainbow was observational rather than theoretical.[17] He was the first person to measure the angular dimensions of the rainbow and establish that the radius of the primary bow is 42°. He also pointed out that each of us sees our own bow, so that there are always as many bows as there

---

[17] Lindberg, D. C. (1966).

are observers and that if the observer is elevated above the ground the bow will be more than a semicircle.

> …it is evident, as we learn from experience, that there are as many rainbows as observers. For if two people stand observing the rainbow in the north, and one moves westward, the rainbow will move parallel to him; if the other observer moves eastward the rainbow will move parallel to him; and if he stands still, the rainbow will remain stationary. It is evident, therefore, that there are as many rainbows as observers, from which it follows that two observers cannot see one and the same rainbow, although an inexperienced person does not comprehend this fact. For the shadow of each observer divides the arc of the rainbow in half; therefore, since the shadows are sensibly parallel, they do not meet at the middle of the same rainbow, and each observer must see his own rainbow.[18]

Renewed interest in the rainbow following the influx of Greek and Arabic works of science, particularly of Alhazen's *Book of Optics*, together with a greater willingness to employ experiments in the search for knowledge and understanding, meant that sooner or later the key to the rainbow would be discovered. In the event, this happened twice within the space of a few years at locations separated by some fifteen hundred kilometres. Inspired by Alhazen's optical theories, at the beginning of the fourteenth century a German monk, Dietrich of Freiberg (died c.1310AD), usually known by his Latinized name, Theodoric, and a Persian mathematician, Kamal al-Din al-Farisi (c.1236–c.1318AD), each performed similar experiments with a spherical flask filled with water and thereby discovered that the rainbow is due to the passage of light through drops of water. Neither man knew of the other, so their discoveries were made independently.

Theodoric was a Dominican monk, a scholar, the author of several books and an administrator of his order; by all accounts a very busy man. As if he didn't have enough to do, in 1304 he was asked by the Master-General of the Dominicans to write an account of his research into the rainbow. The result was *De iride et radialibus impressionibus*, a weighty tome of some two hundred pages setting out in meticulous detail an explanation of the rainbow that in several important respects is not far removed from the one accepted today.

Despite being schooled in the Aristotelian tradition, which you may recall shrank from actively meddling with the natural order it sought to explain, Theodoric boldly set aside his training and performed the one experiment

---

[18] Bacon, R. (1900), p 188.

necessary to get to grips with cause of the rainbow: he filled a spherical flask with water and held it up to the sun, moving it about until he saw flashes of colour at its edge, the same colours that are seen in the rainbow: red, yellow, green and blue. Similar experiments had been carried out by Alhazen and Roger Bacon among others, but Theodoric was the first to realise that these colours were due to light that was reflected from the back of the flask rather than its surface.

> Let [sunlight] enter the [spherical water-filled flask] and pass through it to the opposite surface and from that be reflected internally back to the first surface by which it originally entered, and then passing out let it go to the eye; [this light], I say, inasmuch as it is produced by a transparent spherical body, serves to explain the rainbow.[19]

The experiment enabled Theodoric to establish several important facts about the rainbow. The first of these is that a rainbow is due to light that is refracted and reflected within a raindrop. Secondly, that rainbow colours are seen in a drop only when the angle between the ray of sunlight entering a drop and the one that reaches the eye has a particular value. Here Theodoric had discovered the real reason for the rainbow's distinctive shape: if the angle at which a ray of a particular colour emerges from every drop is always the same, then the drops in which that colour is seen must all lie on the circumference of a circle, the centre of which is at the antisolar point, the point directly opposite the sun on the other side of the sky. Thirdly, that only one colour is seen at a time in any given drop, which means that each of the colours in a rainbow is seen in a different drop to all the others (Fig. 4.5).

Theodoric also explained how the secondary bow is formed. A primary bow, he said, is due to two refractions and one reflection, and a secondary bow to two refractions and two reflections. What's more, he added, because some light is lost each time it is reflected, the secondary bow must be fainter than the primary bow. Furthermore, the dark gap between the primary and secondary bow is due to the fact that no light reaches the eye from drops that lie between the primary and secondary bows because the light responsible for the primary bow emerges from a drop at a different angle than that responsible for the secondary bow (Fig. 4.6).

---

[19] Crombie, A.C. (1953), p 249.

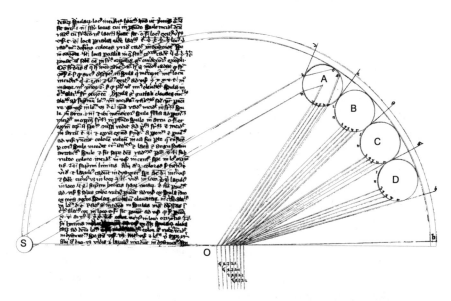

Fig. 4.5 Theodoric's explanation for the primary rainbow, showing how the four colours that Theodoric held were present in the bow drops are produced in drops at different heights above the ground. The sun is at S, the observer at O, and the four drops responsible for colour at A (red), B (yellow), C (green), and D (blue). Sunlight enters the upper half of a raindrop and is reflected once before emerging in the direction of the observer's eye[20]

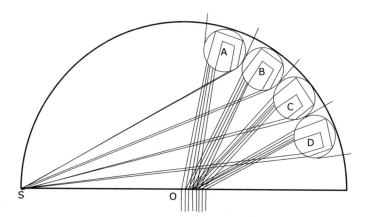

Fig. 4.6 Theodoric's explanation for the secondary rainbow. Sunlight enters the lower half of each raindrop and is reflected twice before emerging in the direction of the observer's eye at O. Colours are seen in A (blue), B (green), C (yellow), and D (red)

[20] Public Domain, https://tinyurl.com/mrzrnbkc Accessed 10/09/22.

But when he took the fruits of his experimental investigations out of the laboratory and into the open air, Theodoric reverted to type and employed Aristotle's sky vault to account for the relative position of the sun, the rainbow and the observer. He drew two diagrams to illustrate the positions of the drops that give rise to the colours seen in a rainbow, one for the primary bow and another for the secondary one. Even allowing for the fact that it is necessary to exaggerate the size of a raindrop in order to show the path of sunlight through it, locating the drops on the surface of the curved sky vault means that that path can't be accurately represented. Theodoric's diagrams add nothing to the insights gained with the glass flask. If anything, they represent a step backwards because they don't represent accurately the actual position of the drops in which the rainbow is seen and misrepresent how it's colours reach the observer's eye.

Nor, for all his independence of mind, was he able to free himself completely from the Aristotelian notion that colour in a rainbow is brought about by the weakening of light, though he parted company with Aristotle on the number of colours. Theodoric claimed that observation established that there were four distinct colours in a rainbow: red, yellow, green and blue. Aristotle might have dismissed yellow as an illusion, but Theodoric was persuaded by the evidence of his own eyes and gave several examples where yellow was always seen distinctly between red and green in drops illuminated by sunlight.

> The same thing is plainly seen also in drops of dew dispersed on the grass if one applies one's eye very close to them so that the drops have a determined position with respect to the sun and the eye. Then in a particular position red appears, but when the eye is moved a little from that position yellow appears plainly and quite distinctly from the other colours. Then with a further change of position the other colours of the rainbow appear in the usual number and order.[21]

Theodoric must have been a skilled and patient experimenter to have got as far as he did with his explanation of the rainbow but he failed to take his investigation to the next step. Having established that refraction and reflection are responsible for the rainbow, he should have made careful measurements that would have enabled him to determine the exact path of sunlight within a spherical drop. But being at heart an Aristotelian, he didn't consider measurement or mathematics to be of relevance to the rainbow, something that is amply confirmed by the diagrams that he drew to show the path of light through a drop. Although these diagrams show clearly the

---

[21] Crombie, A.C. (1953), p 244.

change of direction of light as it enters and leaves the drop, and its reflection within the drop, they don't do so accurately, an indication that he didn't consider measurement an important part of his explanation. His apparent lack of interest in precision is also obvious from his estimate of the size of the primary bow: he claimed that its radius is 22°, rather than 42°.

Unfortunately, Theodoric's work never became widely known, and so the key to explaining the rainbow gathered dust on the shelves of a monastery library while Aristotle's muddled account went largely unchallenged for a further three centuries. Nevertheless, it's almost certain that some details of Theodoric's account remained in circulation because as we shall see in the next chapter, René Descartes, the French natural philosopher and mathematician who came up with the correct explanation for the size of the bow, claimed to have performed an experiment that sounds suspiciously like the one carried out by Theodoric. Descartes was notorious for not acknowledging his intellectual debts, so if he did know of Theodoric's experiments, he would probably not have admitted as much. A full copy of Theodoric's rainbow treatise was eventually discovered in 1813 by Giambatista Venturi, an eighteenth century Italian mathematician and natural philosopher. As we shall see in chapter seven, Venturi made an important contribution to our understanding of the rainbow, though he misinterpreted the results of his experiments to investigate the passage of light through raindrops that are not spherical.

Kamal al-Din al-Farisi's explanation of the rainbow suffered an even worse fate. He was one of the very few Muslim scholars who knew of Alhazen's work on optics, let alone to have studied its contents. Acting on suggestions from his teacher, Qutb al-Din al-Shirazi (1236–1311), concerning Alhazen's experiments with a water-filled glass sphere, al-Farisi made a systematic study of those experiments. He began by making a rigorous mathematical analysis of Alhazen's work, in the course of which he traced the path of rays of sunlight through the sphere even more carefully than had Theodoric. He then confirmed his findings experimentally. Working inside a darkened room into which a narrow beam of sunlight was admitted through a small hole, he placed a spherical glass vessel filled with water in the path of the beam, which enabled him reproduce what happens when a raindrop is illuminated by sunlight. By positioning a white screen near to the vessel on the sunward side he was able to see a faint coloured arc corresponding to that of the primary rainbow (Fig. 4.7). The details of his experiments and the conclusions he reached are broadly the similar to Theodoric's. But Kamal al-Din was the last of his line. No Persian or Arab scholar took up where he had left off. As far as it is possible to tell, with Kamal's death in 1320 AD, interest in the rainbow among Muslim scholars ceased completely.

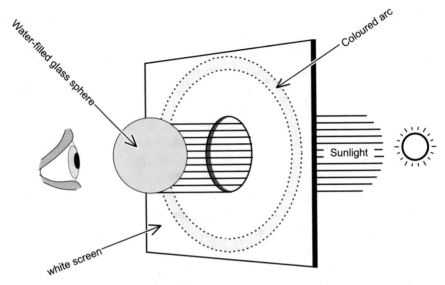

Fig. 4.7 Kamal al Din's experiment to create an artificial rainbow with a sphere of glass. A circular, coloured arc is formed on white screen

That the secret of the rainbow should have been independently discovered more than once is not altogether unexpected. After all, water-filled flasks and crystal spheres had been held up to the sun on countless occasions in the preceding centuries, sometimes for amusement, on others out of curiosity. But it had never occurred to anyone before Theodoric and Kamal that those brilliant flashes of colour seen in a transparent glass sphere filled with of water or in a crystal are the key to the rainbow's secrets. Theodoric and al-Farisi succeeded where others had failed because both of them were familiar with Alhazen's optics and open to his use of experiment to gain insight into natural events. And, of course, they both brought the open-mindedness and imagination that is the hallmark of genius to bear upon the problem.

Their work on the rainbow has been described as the greatest triumph of medieval science, even though it was soon forgotten. In Muslim lands, Islamic scholars turned their backs on the largely secular learning of earlier Arab and Persian philosophers and mathematicians in favour Islamic theology and law, while Europeans of the later Middle Ages, notwithstanding the hostility of the humanist scholars of the Renaissance towards Aristotle, remained largely in thrall to the Aristotelian world view and to his methods for centuries to come.[22]

---

[22] Huff, T. E. (1993), pp 209–39.

And so, although the experiment with the water-filled flask had further important insights to offer, these had to wait until natural philosophers began to marry experiment with mathematics, and that is something that was still a couple of centuries or more in the future.

5

The Geometry of Light

The rainbow is such a remarkable phenomenon of nature, and its cause has been so meticulously sought after by inquiring minds throughout the ages, that I could not choose a more appropriate subject for demonstrating how, with the method I am using, we can arrive at knowledge not possessed by at all by those whose writings are available to us.[1]

On the afternoon of 20th March, 1629, the inhabitants of Rome witnessed a spectacular hour-long display of several unusually bright ice halos, encircling the sun. Halos are not uncommon—if anything, they occur far more often than rainbows—and are formed when the sun is seen through a thin veil of cirrostratus, a type of cloud composed of tiny ice crystals rather than drops of liquid water. What made the *Rome phenomenon*, as it came to be known, unusual was that several distinct halos were seen simultaneously, along with a couple of bright parhelia, or mock suns and a complete parhelic circle. Among those who saw the spectacle was Christoph Scheiner (1573–1650), a Jesuit astronomer famous for his pioneering studies of sunspots.[2] Scheiner wrote a detailed description of the phenomenon that was circulated among

---

[1] Descartes, R. (2001), p 332.

[2] Scheiner constructed the first true astronomical telescope based on a design by Kepler. The eyepiece of a Keplerian telescope employs a convex lens as the eyepiece whereas that of Galileo's telescope was a concave lens (this type of telescope is what we now call an opera glass). The performance of a Galilean telescope is vastly inferior to that of a Keplerian one, which has a wider field of view, is capable of greater magnification and, importantly for observations of sunspots, the image from the telescope can be projected onto a screen. Galileo damaged his eyes by looking at the sun through his telescope.

© The Author(s), under exclusive license to Springer Nature
Switzerland AG 2023
J. Naylor, *The Riddle of the Rainbow*, Copernicus Books,
https://doi.org/10.1007/978-3-031-23908-3_5

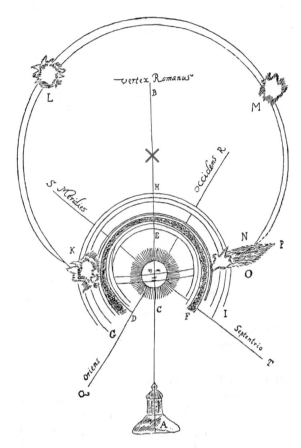

Fig. 5.1 Isaac Beekman's sketch of the *Rome Phenomenon*. KLMN is a colourless parhelic circle that is centred on the zenith (marked with an X in the image). The observer would be at A[4]

natural philosophers in France and beyond (Fig. 5.1). A few months later, news of the event reached René Descartes (1595–1650), an ambitious French mathematician destined to become a mechanical philosopher *par excellence*.[3]

Descartes was 33 years old at the time and had yet to make his name, though he was acquiring a reputation as an up-and-coming savant, if only in France. He had already made major discoveries in mathematics and optics but had yet to publish anything. He had gone to live in Holland the previous year to escape the distractions of Paris and in search of somewhere to pursue

---

[3] Tape, W., Seidenfaden, E., Können, G. P. (2008), p 76.

[4] Isaac Beeckman, Journal, Cornelis de Waard, ed., 4 Volumes, The Hague: Martinus Nijhoff, 1939–45, Vol IV, p 150.

his studies in peace. Having inherited money from his mother, he had no need to earn a living and was free to follow his interests.

Descartes was born into a well-to-do family in 1596 in La Haye—renamed Descartes in his honour in 1967—a town on the river Creuse, some 25 km south of Tours. At the age of ten he was sent to board at the Jesuit college of La Flèche at Anjou, considered at the time to be one of the best schools in France. He was a frail child, and according to uncorroborated accounts was provided with a private room and allowed to remain in bed well beyond the hour at which the rest of the school had to rise. Whatever the truth of this, as an adult he gained a reputation for late rising—he once confessed that he regularly slept for "ten hours every night"[5]—and was known to prefer his bed to a desk as a place of work. What little else is known for certain of his school days comes from a few biographical remarks in the preface to *Discourse on Method*, his first published work.[6] Reading between the lines, it seems that he was an able, thoughtful and conscientious pupil and that he enjoyed the years he spent at the school—he remained in contact with his teachers throughout his life—but that his schooling left him with an abiding scepticism of ideas founded on authority alone.

He left La Fléche in 1614 and headed for Poitiers, where he studied law at the university. He received his degree in 1616. The following year he made his way to the Netherlands where he joined the army of Maurice of Nassau, Prince of Orange (1567–1625), which was fighting the Spanish forces of the Hapsburg Empire. Descartes never saw active service but a chance meeting in the autumn of 1618 in the Dutch town of Breda with Isaac Beekman (1588–1637), a Dutchman with broad scientific interests, rekindled his earlier interest in science and mathematics. Beekman introduced Descartes to the latest scientific ideas and, by challenging him to come up with mathematical solutions to problems in mechanics, hydrostatics and acoustics, was instrumental in developing Descartes' latent mathematical talent.

But according to Descartes, the defining event in his life occurred in the winter of 1619 when a series of vivid dreams during the course of a single night convinced him that he should "…devote my whole life to cultivating my reason and advancing as far as I could in the knowledge of the truth".[7] For several years following this epiphany he travelled widely in Holland, Germany, Italy and France until late in 1628, when he decided to settle in Holland. There he lived a largely quiet and secluded life, corresponding

---

[5] Descartes, R. (1897), pp 198–9.
[6] Descartes, R. (2001).
[7] Descartes, R. (2001), p 23.

with leading thinkers of the day and writing several of the key works of the mechanical philosophy that was to sweep aside the prevailing Aristotelian world-view during the latter half of the seventeenth century. The central tenet of this new philosophy was that nature is a vast, inanimate mechanism, and the most complete statement of the physics and cosmology that flowed from this idea is to be found his *Principia Philosophiae* (The Principles of Philosophy), published in 1644. It quickly became the bible of mechanical philosophy and gave the sciences, particularly the mechanical sciences, a much-needed shot in the arm that led to spectacular advances in the sciences of dynamics and optics later in the century.

In 1649, in the mistaken belief that a prince has much to learn from a philosopher and seduced by promises of a title and a pension, Descartes rashly accepted an invitation to travel to Sweden and tutor the young and capricious Queen Christina of Sweden (1626–89). Required to attend her several times a week in her quarters at 5 am during a bitter northern winter, his frail constitution let him down; he caught pneumonia and died 11th February, 1650. Christina was so shocked by his death that she decreed he be given a state funeral and interred in a marble tomb in the royal cemetery. Meanwhile, until this could be arranged, as Catholic in a Lutheran state, he was buried in a graveyard reserved for children who had died before they could be baptised. But Christina abdicated before she made good her promise and Descartes' body remained in its temporary grave until reclaimed by the French seventeen years later. Ironically for a man whose reputation rests on his intellect, only his body is buried in Paris; the exact whereabouts of his skull is not known.[8]

Although Descartes had dabbled in optics following his meeting with Beekman in Breda, his interested in the subject began in earnest while living in Paris from 1625 to 1628. At the time, the talk among French natural philosophers was of ways to improve the telescope, then in its infancy as a scientific instrument, and Descartes soon found himself swept up by the widespread interest in optical matters. Over the next three years he devoted much of his time to the search for Holy Grail of optics, the law of refraction, in which he succeeded, and to the application of this law in the design of the perfect telescopic lens, in which he failed.

The law of refraction proved to be the key to the explanation of the rainbow, so when a letter that included an account of the Rome phenomenon came Descartes' way in the summer of 1629, it had reached the only person in the world at that time with the wherewithal to do more than merely marvel at

---

[8] Grayling, A.C. (2005), p 272.

Scheiner's description of the spectacle of several eye-catching luminous arcs. Descartes was particularly intrigued by the apparent similarity between the vivid halos seen in Rome and the rainbow: both are narrow circular arcs consisting of concentric coloured bands that appear only when the sun is visible and clouds are present.[9]

At the time, the cause of ice halos and parhelia was even more of a mystery than that of the rainbow, but Descartes was confident that he could come up with an explanation for all three phenomena.[10] That autumn, he wrote to his friend and confidant Marin Mersenne (1588–1648), a French monk and an accomplished mathematician and natural philosopher, that he had resolved "to write a small treatise which will contain the explanation of the colours of the rainbow".[11] In fact, Descartes went on to say, the treatise would deal with all sublunary phenomena, not just the rainbow. Convinced since his epiphany in 1619 that science and philosophy were in need of root and branch reform and believing that the only way to do this was to tear up the existing rule book and begin anew, Descartes considered his treatise as a means to an end. He begged Mersenne to say nothing about the project to anyone because "I have decided to exhibit it publicly as a sample of my Philosophy, and to hide behind the canvas to listen what people will say about it."[12]

Although Mersenne was not in the same league as Descartes, either as a mathematician or a philosopher, he was one of the most inventive and skilled experimentalists of his day, easily the equal in that respect to any of the savants of that era, including Galileo Galilei (1564–1642). He also kept in close touch with most of the leading natural philosophers and mathematicians of the time, his address book a who's who of European natural philosophers and mathematicians during the first half of the seventeenth century. Indeed, he was Descartes' chief avenue of communication with the learned world. He also organised regular meetings of savants in his rooms in Paris—meetings that eventually led to the creation of the state-funded Académie Royale des Sciences, set up in 1666 by Jean Baptiste Colbert (1619–83), chief minister to France's formidable monarch, Louis XIV (1638–1715), Le Roi Soleil[13]—and acted as a clearing-house for the latest scientific ideas and discoveries by circulating copies of letters he received from his

---

[9] Some ice halos, such as the so-called 22° halo, are often colourless. But several other ice halos are as colourful as the rainbow, parhelia and circumzenithal arcs particularly so.

[10] Ice halos exist in so many different forms that their diverse causes were not satisfactorily established until the nineteenth century by Auguste Bravais in 1845. See: Bravais, M.A. (1845), pp 77–96.

[11] Descartes, R. (1897), p 23.

[12] Descartes, R. (1897), p 23.

[13] *Le Roi Soleil* = The Sun King.

numerous correspondents.[14] He was one of the first savants to recognise the importance of Galileo's scientific work and wrote to him offering to help with the publication of "the new system of the motion of the earth which you have perfected, but which you cannot publish because of the prohibition of the Inquisition."[15] Bearing in mind that Mersenne was a devout Catholic and an ordained priest, his offer to make Galileo's ideas better known despite the prohibition issued by the Inquisition is surprising.

On the face of it, Mersenne may seem to be a most unlikely recruit to the ranks of mechanical philosophers. But he became convinced that the new mechanical philosophy could help the Catholic Church combat the occult sciences such as magic, astrology and alchemy that had flourished in Europe since the early Renaissance and which blurred the line between the natural and supernatural world.[16] Mersenne believed that the inert, inanimate material world that underpinned mechanical philosophy allowed for the possibility of miracles, i.e. events that are of out of the ordinary and which could be explained only as the work of God. The practitioners of the occult sciences claimed that they could achieve similar marvels because the material world was a living organism and that they alone understood its deepest secrets and how to manipulate them.[17]

Over the next few months Descartes kept Mersenne abreast of progress, though he continued to press him to keep things under his hat. Descartes was notoriously secretive, not to say paranoid—it is more than likely that had he been alive today, he would have avoided using a phone and used a post-office box for his mail—so his desire to make his ideas known to the world while remaining anonymous wouldn't have come as a surprise anyone who had had dealings with him. The project rapidly grew in scope. A few weeks later he wrote again to Mersenne and announced that "instead of explaining a single phenomenon, I have decided to explain all natural phenomena, that is, the whole of physics."[18] Thus from tiny acorns do mighty oaks grow!

It didn't take Descartes more than a few months to come up with an explanation of the rainbow, even though, as we shall see, it was later shown to be

---

[14] How Mersenne found the time and energy for these extracurricular activities is astonishing because the religious order to which he belonged, the Minims, was one of the most ascetic orders in France. In addition to the daily round of ritual observances, Minims undertook a perpetual Lenten fast, i.e. they avoided eating meat and dairy products and went barefoot.

[15] This was Galileo's *Dialogue Concerning the Two Chief World Systems*, which was eventually published in 1632.

[16] Grayling, A.C. (2016), p 119.

[17] The ability to manipulate nature is known as *natural magic*. Alchemy is an example of natural magic.

[18] Descartes, R. (1897), p 70.

incomplete even within the terms he had set himself, but it was another eight years before the projected work saw the light of day in 1637 under the title *Discourse on Method, Optics, Geometry and Meteorology*.[19]

The delay was due to news that Galileo, Europe's foremost natural philosopher of the time, celebrated for his telescopic discoveries, had been brought before the court of the Inquisition in Rome, accused of breaching an injunction that had been imposed on him by the Church in 1616 not to teach the Copernican doctrine that the earth orbits the sun. The ban had been engineered by Galileo's Aristotelian rivals in the universities rather than the theologians of the Church, who nevertheless took the philosophers' side. The gist of the ban was that the Copernican doctrine, for which Galileo had argued in several of his publications, was contrary to Holy Scripture, and that it was not possible to argue in its favour using arguments based on natural philosophy—the issue could be settled only by means of philosophical reasoning, something that suited the Aristotelians down to the ground.[20]

In 1632, Galileo published *Dialogue Concerning Two Chief World Systems* in which the Copernican doctrine was clearly favoured over the geocentric world system of Aristotle. The ban was retrieved from the archives, probably by Christoph Scheiner who had earlier crossed swords with Galileo on the question of the nature of sunspots, and Galileo was summoned to Rome to face the Inquisition. He was found guilty of breaking the ban, put under house arrest and all available copies his book burned.

When news of this reached Descartes, he panicked because his own book, provisionally entitled Le Monde, was openly Copernican, much of it given over to an elaborate physical explanation for orbital motion of planets about the Sun. He immediately began to disown the manuscript and distance himself from its contents. He wrote to Mersenne "I have never had the inclination to produce books, and would never have completed it if I had not been bound by a promise to you and some of my other friends..."[21] A couple of months later he was even prepared to admit that his defence of the Copernican system was flawed: "Though I thought that they were based on very certain and evident proofs, I would not wish, for anything in the world, to maintain them against the authority of the Church."[22]

Descartes' alarm was unnecessary because as a resident of the Dutch Republic he was well beyond the reach of the Roman Inquisition. Nevertheless, he shelved the manuscript of Le Monde and it was not published in

---

[19] Descartes, R. (2001).
[20] Drake, S. (1980), pp 61–5.
[21] Descartes, R. (1897), p 250.
[22] Descartes, R. (1897), p 282.

his lifetime.[23] However, despite protesting to Mersenne that he had never wished to write a book, Descartes decided to cannibalise its less controversial sections, the ones that dealt with light and meteorology, and publish them as self-contained essays. And in a change of tack, he decided to preface them with a philosophical essay, to which he gave the title *Discourse on Method*, which would meet the Inquisition's ruling that ideas in natural philosophy can only be supported on metaphysical grounds.[24]

True to his word, the author's name appeared nowhere in the book. Moreover, despite the fact that in seventeenth century Europe Latin was the international language of science and philosophy, he wrote it entirely in French because the readership he had in mind was the educated public of France rather than the Aristotelian pedants of the universities, whom he regarded as a lost cause.

> And if I write in French, which is the language of my country, in preference to Latin, which is that of my teachers, it is because I believe that those who make use of their unprejudiced natural reason will be better judges of my opinions than those who heed only the writings of the ancients; and as for those who unite good sense with habits of study, whom alone I desire for judges, they will not, I feel sure, be so partial to Latin as to refuse to listen to my arguments merely because I expound them in the vulgar tongue.[25]

The initial print run was three thousand copies, of which Descartes received two hundred for his own use. He had high hopes for it. Unfazed by his declared wish to remain anonymous, he sent several copies to his old school in the hope that it would be adopted as a textbook (it wasn't) and distributed the remainder to Jesuits, diplomats and leading lights of Parisian society. He even sent a copy to Cardinal Richelieu, who Descartes knew was interested in reforming French education—though what that arch political manipulator made of the book is not known. In the event, the book was not the instant success he had hoped it would be; at least, not in the French edition. Thirteen years after its publication, in 1650, the year of Descartes' death, his publisher still had unsold copies on his shelves. Hardly surprising given that Descartes had probably exhausted the initial market for the book by distributing a couple of hundred free copies. It was only after it was translated into Latin in 1644 that the *Discourse* and its attendant scientific essays

---

[23] It was eventually published in Paris in 1664 under the title *Le Monde de M. Descartes ou le Traité de la Lumierè.*

[24] Gaukroger, S. (1995), p 321.

[25] Descartes, R. (2001), p 62.

became widely known outside France. The mathematical essay, the *Geometry*, was not translated into Latin until 1649.

Confident that his readers would be won over by the ideas and explanations contained in the essays, Descartes invited anyone "who may have any objections to make to them, to take the trouble of forwarding these to my publisher, who will give me notice of them, that I may endeavour to add at the same time my reply."[26] But those who took the offer up found that the gracious selflessness of this invitation evaporated in the face of criticism. Instead of answering his critics, Descartes rubbished them; he was unwilling to make any concessions or admit any errors. Letters from one doubter were fit only as "toilet paper". Other critics were dismissed as "a little dog", "a less than rational animal", and several were mere "flies". As for the reservations expressed by Pierre Fermat (1601–65), by common consent a greater mathematician than Descartes, what could one expect from a man whose mathematical work was a "pile of shit"?[27]

Discourse on Method is a lengthy philosophical tract in which Descartes sets out and justifies a method of reasoning of which other three sections of the work, Optics, Meteorology and Geometry, are examples of its application. Optics expounds his ideas about the nature and properties of light. Among other things, it includes the first published account of the law of refraction. Meteorology is an attempt to update ideas held by the ancient Greeks, particularly those of Aristotle, about weather-related events such as winds and clouds using the principles of mechanical philosophy. This is where Descartes includes his explanation of the rainbow, ice halos and parhelia. The third treatise, Geometry, is a demonstration of how algebraic methods can be applied to geometric problems.[28]

The full title of the philosophical preface is *Discourse on the Method for Rightly Directing One's Reason And Searching for the Truth in the Sciences*. It begins as an intellectual autobiography: "my intention is not to teach here the method which everyone must follow in order to direct his reason correctly, but only to show the manner in which I have tried to direct mine."[29] He tells us that at school he learned many useful things: languages, poetry and, above all "the certitude and clarity" of mathematics. But he found philosophy and, by extension, science, disappointing "considering how many diverse opinions about a single subject there are in philosophy, and that these are upheld by learned men, although there can never be more than one among them that is

---

[26] Descartes, R. (2001), p 60.
[27] Shea, W.R. (1991), n 37, p 292.
[28] Known as *Analytic* or *Cartesian* geometry.
[29] Descartes, R. (2001), p 5.

correct, I deemed everything which was merely plausible to be almost false."[30] If the teachings of established authorities can't be relied upon, how is one to distinguish truth from falsehood? What is needed is an infallible method that will enable one to determine what is true, a method which he, Descartes, had discovered.

The heart of this method seems unexceptionable and self-evident. It is "never to accept anything as true that I did not know evidently to be such".[31] But the criterion of Cartesian truth is problematic, based as it is on "that which presented itself to my mind so clearly and so distinctly that I had no occasion to place it in doubt".[32] What, after all, is the hallmark of undubitable ideas? Putting himself to the test, Descartes claimed that he could call into question just about everything that he ordinarily held to be certain. He could doubt that the world existed (it might be no more than a vivid dream) and doubt he had a body (it could be a delusion). Even the certainties of arithmetic are not to be trusted: "...how do I know that I am not deceived every time that I add two and three, or count the sides of a square...".[33] The only thing one can be sure about is that there are doubts, which, he says, presuppose a doubting being. In other words, the one thing that cannot be doubted is one's own existence, if only as a collection of thoughts. Hence the most notorious sound bite in the history of philosophy, "I think, therefore I am" (it's even snappier in Latin: "cogito, ergo sum"). The point of the exercise was not, however, to arrive at a paralysing scepticism, it was to establish a bedrock on which he could erect an edifice of irrefutable knowledge about the world; and in "I think, therefore I am", he believed he had found it.[34]

But of what use is this one certainty if it is possible to comprehensively demolish our faith in everything that we ordinarily take to be certain or true? Descartes had an answer, though it seems little more than a sticking plaster, and not a very adequate one at that. There was one idea, he said, of which he was unable to divest himself through methodical doubt: that of God, the very embodiment of perfection. Descartes argued that because doubt suggests incompleteness, which he claimed implies a lack of perfection, the source of the idea of a perfect being must exist independently of a doubting mind. Ergo, God, the perfect being, must exist.[35] Moreover, such a

---

[30] Descartes, R. (2001), p 8–9.

[31] Descartes, R. (2001), p 16.

[32] Descartes, R. (2001), p 16.

[33] Descartes, R. (1996), pp 58–63.

[34] Descartes, R. (2001), p 27.

[35] Descartes' proof for the existence of God is a variation of St Anselm's Ontological Argument, c 1100 AD.

being is necessarily benign and so would not deceive us without good reason. Descartes thus makes God the ultimate guarantor of the truth of ideas that are "clearly and distinctly" perceived. A cynic might point out that, rather conveniently, Descartes' god was the Christian god of love rather than, say, Zeus, the supreme god of the ancient Greeks. Zeus had no compunction about deceiving human beings when it suited him. In order to seduce Leda, wife of Tyndareus, king of the Spartans, he approached her in the guise of a swan being pursued by an eagle so that she would allow him to lie with her.

These ideas about the foundations of knowledge, which he elaborated and refined in later books, made Descartes' reputation as a philosopher, but they were his undoing as a scientist. He acknowledged as much when he admitted "The things that we conceive very clearly and very distinctly are all true, but…there only remains some difficulty in properly discerning which are the ones that we distinctly conceive."[36] An understatement, if ever there was one.

The weakness of Descartes' scientific method is his belief that the only way to establish true knowledge about the world is through the exercise of reason alone. He attached little importance to hands-on investigation, and never accepted that a scientific theory can be brought to its knees by empirical facts. Consequently, experiments for him always played second fiddle, fit only to check the finer points of explanations arrived at through logical deductions from general principles that were themselves "clearly and distinctly" perceived to be true. As a quick glance through his work on meteorology shows, he had little interest in making new discoveries and confined himself to explanations of familiar phenomena, many of which are taken directly from Aristotle's Meteorologica. A notorious example of the nonsense that all too often resulted from the combination of the blithe confidence he had in his method and his cavalier approach to establishing the facts before attempting to account for them is his breezy explanation of how sometimes it can rain "iron, blood or locusts or similar things".[37]

Radical doubt was important to Descartes for another reason. He believed it establishes that reality consists of two distinct realms: mind and matter. The essence of mind is disembodied thought and because matter occupies space its essence is extension. Furthermore, he declared, matter is infinitely divisible and thus can be made as small as necessary to fill even the tiniest space. As a result, in a Cartesian universe nowhere is free of matter and consequently Descartes denied that a vacuum is possible. Moreover, because every fragment of matter is in direct contact with its neighbours, in the material world everything that happens in it does so because when one bit of matter moves

---

[36] Descartes, R. (2001), p 28.
[37] Descartes, R. (2001), p 339.

it pushes against its immediate neighbours. Minds, however, are immaterial, so our knowledge of the material world is necessarily indirect. The colours, smells and sounds that common sense assigns to the world are, according to Descartes mental events brought about when matter impinges on our nervous system. The material world itself is colourless, odourless and silent, quite unlike the one we see, smell, hear or feel. But how the mind is able to communicate with the material world and vice versa, the so-called mind–body problem, was something that Descartes never solved satisfactorily, and one that continues to vex philosophers and neuroscientists to this day.

Surprisingly, having ploughed through these arguments to establish a method for discovering truth, Descartes made no appeal to it in any of the examples of his scientific discoveries and explanations that accompany the *Discourse*, except in his explanation of the rainbow, which forms the eighth chapter of the essay on meteorology. He begins by pointing out that the "rainbow is such a remarkable phenomenon of nature, and its cause has been so meticulously sought after by enquiring minds through the ages, that I could not choose a more appropriate subject for demonstrating how, with the method I am using, we can arrive at knowledge not possessed at all by those whose writings are available to us."[38]

But reading through his explanation of the rainbow one soon finds that it is not quite the cerebral exercise that his avowed confidence in the importance of clear and distinct ideas would lead one to expect. Instead, Descartes relies on experimental evidence and mathematical analysis, the twin pillars of modern science. Indeed, the explanation of the rainbow is the only example in his entire essay on meteorology that we would nowadays recognise and accept as science. All the other meteorological essays are little more than unconvincing scientific fantasies, the products of speculation unrestrained by experiment.

At the time of their publication in 1637, the scientific essays aroused more interest than the philosophical preface. But Cartesian science long ago failed to live up to its promise and today the situation is completely reversed. *The Discourse on Method* is studied in minute detail, if only by philosophers attracted like moths to a flame by the conundrum posed by Cartesian doubt and the resulting paradox of how an immaterial mind can know anything of a material world, while his scientific work is largely forgotten because, except for the law of refraction and the explanation of the rainbow, most of Descartes' scientific theories and explanations have not stood the test of time. That is not to say that some of his ideas were not influential. His principle of

---

[38] Descartes, R. (2001), p 332.

inertia, the idea that in the absence force a body will move in a straight line at constant speed, was later adopted by Isaac Newton.[39] But within a couple decades of Descartes' death in 1650 Newton exposed Cartesian cosmology as wholly mistaken.

Descartes' distrust of established ideas, which to judge from his auto-biographical remarks he developed while still at school, together with an unshakable confidence in his intellectual powers, often led him either to ignore or to disparage the ideas of others. Reading through his essay on the rainbow one would never guess that several of his contemporaries had come close to the correct explanation. Setting aside the fundamentally flawed Aristotelian account that had held sway for two thousand years, which Descartes could justifiably ignore as a non-starter, most natural philosophers of the time accepted that a rainbow is due to what happens when light passes through a drop of water. But, for one reason or another, their explanations all failed to exploit this insight.

Francesco Maurolico (1494–1575), a Benedictine monk and mathematician from Sicily, realised that an important issue that any explanation of the rainbow must address is that of the apparent size of its two arcs. Brushing aside the irksome task of actually performing any experiments, he opted instead for a plausible theory based on repeated reflections of sunlight within a spherical drop and ignored the role of refraction completely. Another leading mathematician, the first natural philosopher to go beyond Alhazen's optics and lay the foundations of modern optical theory, Johannes Kepler (1571–1630), better known to posterity for his discovery of the laws of planetary motion, realised that both refraction and reflection were involved. But he assumed wrongly that a rainbow is due to tangential rays that just graze the surface of the drop. Like Maurolico, Kepler also avoided performing experiments where possible. Mesmerised by mathematics, both men came to the conclusion that the true radius of a rainbow must be 45°, or half a right angle. Neither of them was prepared to tackle squarely the awkward fact that its radius is actually 42°, a fact that had been established by Roger Bacon some four centuries earlier.[40] Kepler maintained that the larger value suited his explanation and waved aside the difference by claiming that the water in a raindrop is probably tepid, making it less refractive than normal. Maurolico attributed the difference to the possibility that in nature raindrops are not perfectly spherical, which as we shall see in Chap. 7 is indeed true for drops larger than 1 mm (Fig. 5.2).

---

[39] This is "Newton's First Law of Motion".
[40] Kepler, J. (2000), p 168.

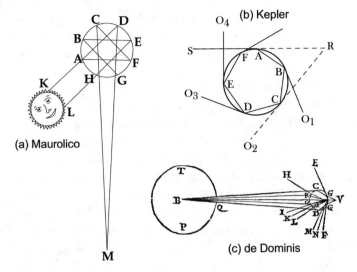

Fig. 5.2 The path of the ray of light through a raindrop responsible for a rainbow according to **a** Maurolico, **b** Kepler, **c** de Dominis

Maurolico's explanation of the rainbow was published posthumously in 1611. Remarkably, another account of the rainbow appeared that year, the posthumous work of Marco Antonio de Dominis (1564–1624), the infamous Bishop of Spalato,[41] said by J. W. Goethe (1749–1832) to have discovered the spectrum while saying mass. Dominis began life as Jesuit and taught mathematics and philosophy at the University of Padua. He was made a bishop in 1600 but converted to Protestantism in 1616 after falling foul of Pope Paul V (1550–1621) for his criticism of the Catholic Church—the same pope responsible for the injunction that forbade Galileo from teaching the Copernican doctrine. Dominis sought sanctuary in England and was appointed Dean of Windsor by the King, James I (1566–1625). However, his insatiable avarice and vanity eventually alienated his English friends and in 1622, following Paul V's death, he reverted to Catholicism and returned to Rome. Predictably, he was arrested by the Inquisition, charged with apostasy and condemned to the stake. He died before the sentence could be carried out. The Inquisition was not to be thwarted: his body was exhumed and burned, along with his all his writings.

At first glance, Dominis' explanation of the rainbow appears similar to Theodoric's, from whom he may well have borrowed the idea that sunlight refracts on entering the drop and some of it is reflected from the rear surface of the drop towards the observer. But unlike Theodoric, Dominis seems not

---

41 Spalato was the Italian name for Split, in Croatia.

to have realised that light undergoes a second refraction as it emerges from the drop on its way to the eye. Nor did he realise that the secondary bow is due to rays that undergo a second reflection within the drop. Nevertheless, as we shall see in the next chapter, on the subject of the rainbow, Descartes' achievement was to be dogged by Dominis's flawed account of the rainbow well into the twentieth century.

As the foremost theoretician on optics of his day, Kepler had a long-standing interest in the rainbow, though he published little on the subject; most of his ideas about rainbows are to be found in his letters. In fact, in 1606 he was briefly in touch with someone who had already arrived at the correct explanation the rainbow and its colours.[42] This was Thomas Harriot (1560–1621), a remarkably versatile and original English mathematician, astronomer and natural philosopher. Harriot's patron was the adventurer and courtier Sir Walter Raleigh (1552–1618), to whom he acted as a scientific advisor. Harriot had sailed to North America in 1585 with an expedition organised by Raleigh, the intention of which was to create an English settlement on the coast of what later became the state of Virginia. During his stay Harriot studied the customs of the local Algonquian people, having already learned their language in London from two Algonquians who had been (willingly) taken to England by an earlier expedition. Prior to the voyage, Harriot had devised novel mathematical methods of navigation to aid the captains of Raleigh's ships. On his return to England he wrote an account of the New World and its inhabitants.[43] The settlement itself did not prosper, and when a much-delayed supply ship arrived in 1590 its inhabitants and the buildings had vanished without trace.

Harriot's other patron was Henry Percy, 9th Earl of Northumberland (1564–1632). Percy was nicknamed "The Wizard Earl" due to his interest in mathematics and natural philosophy. In 1597 he installed Harriot in a house at Syon Park, his estate on the Thames upriver from London and gave him a generous stipend and free reign to pursue his many interests.

Among his many researches at Syon, in 1609, Harriot became the first person to study the moon through a telescope, stealing a march of several months on Galileo.[44] The following year he plotted the progress of sunspots across the face of the Sun and measured the orbital periods of the moons of Jupiter.[45] But one of the first topics that Harriot worked on at Syon was

---

[42] Kepler, J. (1859), pp 67–72.

[43] Harriot, T. (1590).

[44] Harriot's best maps of the moon are superior to those of Galileo, though not as aesthetically pleasing. Compare images of "Harriot's Moon" and "Galileo's Moon": https://tinyurl.com/34rd27ax. Accessed 8/03/2022.

[45] Arianrhod, R. (2019), pp 214–21.

optics. He made careful measurements of the refraction of light from air to water that were both systematic and accurate. These led him to the law of refraction in its modern mathematical form, probably as early as 1601, a quarter of a century before Descartes.[46] Five years later Harriot used this law to determine the precise path of several rays of light through an idealised spherical drop of water and thus discovered why the radius of a rainbow is 42°. He was even able to show that the colours in a rainbow were due to dispersion because his measurements of refraction when light passes from air to water were sufficiently precise that he discovered that refractive index depends on colour, with red being the least refracted colour.[47] Unfortunately, he never published any of his work in optics and said almost nothing in the course of his brief correspondence with Kepler on the subject of the rainbow that Kepler did not already know. Harriot's reticence probably owed much to Kepler's brazen demand that Harriot share all his measurements and ideas on refraction and rainbows. Hence Kepler never learned that Harriot had discovered the law of refraction, which Kepler knew was the key to optics and for which he had sought in vain. So, despite this correspondence, unlike Theodoric's account of the rainbow, Harriot's work on optics and on the rainbow remained completely unknown to the natural philosophers and mathematicians of the early seventeenth century who might have profited from his insights.

Unfortunately, by not publishing during his life he took this knowledge to the grave. Too late, he realised his mistake. As he lay on his deathbed—dying from a cancer of the nose, possibly brought about by his habit of "drinking smoke" that he had acquired during his stay in America thirty-five years earlier—he begged his friends to edit and publish his life's work. But for a variety of reasons this never happened and Harriot's research papers, like those of Theodoric, gathered dust until an Austrian astronomer, Franz Xaver Zach (1754–1832), stumbled across them in 1784 under a pile of stable accounts in the library of Petworth House, a stately home in South East England.[48] One of the reasons why there was no posthumous edition of Harriot's papers was that he was an atomist, a doctrine that at the time was considered dangerously atheistic and with which few, if any, of his contemporaries would have wished to be openly associated.[49] In any case, Harriot

---

[46] The law of refraction is: $sin\ i = k\ sin\ r$, where i and r are the angles of incidence and refraction respectively, and k is the refractive index. Harriot used his own notation for trigonometrical quantities.

[47] Arianrhod, R. (2019), chaps 12 and 16.

[48] Arianrhod, R. (2019), pp 261–3.

[49] Harriot's papers have been digitized and are available online: https://tinyurl.com/brk8ejyr. Accessed 2/08/2022.

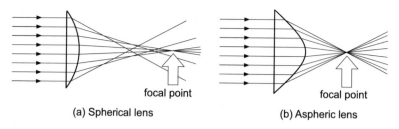

focal point

(a) Spherical lens

focal point

(b) Aspheric lens

Fig. 5.3 Rays of light through **a** a spherical plano-convex lens **b** an aspheric plano-convex lens

was without peer; none of his friends or colleagues were up to the task of sifting systematically through the huge number of bundles of his notes and organising the insights and discoveries contained in them into a publishable form.

True to form, Descartes never revealed exactly what led him to realize that he could explain the rainbow, but we can hazard a guess. He had discovered the law of refraction two or three years before the description of the Rome phenomenon reached him, but had kept it to himself. As he thumbed through Scheiner's account it must have dawned on him that in this law he had the key to the rainbow. He had already used the law to determine several possible profiles for aspheric lenses in 1627, the very type of lens that the Parisian natural philosophers had sought in vain when Descartes' was living among them. In the right circumstances, it was believed, such a lens would focus all parallel rays to a single point, something that a lens with spherical surfaces can never do (Fig. 5.3).[50]

Descartes hoped that his designs for aspheric lenses would enable opticians to improve the telescope by overcoming one of the major problems that bedevilled all lenses until well into the eighteenth century: their inability to focus a broad beam of light precisely, thereby forming a fuzzy image. To establish the exact shape of such a lens, Descartes had used the law of refraction to determine the path of several imaginary rays of light through a lens, a technique now known as ray-tracing.[51] In fact, despite Descartes' claim to the contrary, a Cartesian aspheric lens was never made because it called for techniques that were beyond the skills of even the best lens makers of the

---

[50] An aspheric lens can't overcome chromatic aberration due to the fact that white light is a combination of colours of different refractive index. But it is possible to do so by combining two lenses of different types of glass. The invention of the so-called achromatic lens is credited to Chester Moore Hall, an English lawyer and inventor, in 1733.

[51] Recently, scholars have found that an aspheric lens using a form of the law of refraction was designed by an Arab mathematician, Ibn Sahl (940–1000), possibly in 980 AD. See: Rashed, R. (1990).

time. Nevertheless, having employed ray-tracing in the design of the ideal lens, he would have realised that he could use the same technique to explain the rainbow. It would be nice to think that at that moment he leapt from his bed with a cry of "Eureka", like a latter-day Archimedes.[52]

Surprisingly, it isn't necessary to know anything about the nature of light in order to explain many of the rainbow's features. But to succeed in this it is necessary to think about light in terms of rays, fictitious lines that show the path light takes as it travels through a medium. It's also necessary to know the exact amount by which this path changes direction as light is reflected or refracted by a drop of water, information that can be calculated precisely using the laws of reflection and refraction.

The convention of representing light as rays and the law of reflection had been established in antiquity by Euclid (c.300 BC), the Greek mathematician and author of the Elements, one of the most important and influential works on geometry ever written and among the most improbable best-sellers of all time—it is likely that more copies of the Elements have been printed than any other book except for the Bible.[53] Euclid's law of reflection, which is as valid today as it was when he discovered it, states that the angle at which light is reflected is always the same as the angle at which it impinges on the reflecting surface: the angle of incidence equals the angle of reflection—yet another fact you may have learned at school. But, because reflecting surfaces are seldom perfectly flat, these angles are always measured from a line known as the normal, which is perpendicular to the surface at the point of reflection, rather than from the surface itself (Fig. 5.4).

The law of refraction, however, proved altogether more elusive because when light changes direction as it passes from one transparent medium to another the angle of refraction is never the same as the angle of incidence. We now know why: refraction is a property of waves and occurs when a wave changes speed. The circumstance in which this happens varies from one type of wave to another. The speed of waves travelling across the surface of water depend on depth, being less in shallow water than in deep water; sound travels faster in warmer air; and light slows when it passes from a rarefied transparent medium such as air to denser ones such as water or glass.

In the case of light, the change in speed—and thus the amount of refraction—can be calculated from the so-called refractive index of the material.

---

[52] Archimedes is said to have run naked through the streets of Syracuse after realising that he had discovered how to determine the purity of the gold from which the Tyrant's crown was made without destroying it.

[53] Euclid's *Elements* was unknown in Western Europe until translated by Adelard of Bath (c1180–c1152) in 1120 from a copy in Arabic.

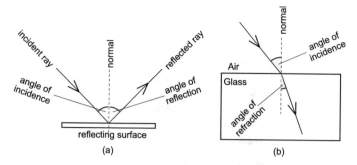

Fig. 5.4 Ray diagrams illustrating the path of a ray of light when it is **a** reflected and **b** refracted

Refractive index is the ratio of the speed of light in a vacuum to the speed of light in the material. In the case of water this is approximately 1.5 because the speed of light in water is two-thirds the speed of light in a vacuum. The degree of refraction also depends on the frequency of the light, the number of times it vibrates in a second, which our eyes perceive as colour, blue having a frequency approximately twice that of red. The refractive index of violet light in water is roughly 1% greater than that of red light, just enough to prise them apart sufficiently for the eye to see each colour distinctly under the right circumstances.

Examples of the refraction of light are ubiquitous. Refraction makes the depth of water in a swimming pool or a bowl seem less than it really is. Refraction is also the reason why the handle of a teaspoon or the shaft of a pencil, partially immersed in water, appears sharply bent where it enters the water. And because the degree of refraction depends on frequency it some-times gives rise to colours, such as at the bevelled edge of a thick sheet of glass, in a cut gem and, of course, in a drop of water.

Students of optics have for generations been introduced to the phenomenon of refraction through an experiment attributed to Euclid and which is among the oldest experimental demonstrations of a scientific prin-ciple. A coin is placed in an empty, opaque vessel, say a cup, in such a position that the coin is just hidden from sight by the vessel's rim. When the vessel is slowly filled with water, the coin gradually comes back into view (Fig. 5.5).

This simple demonstration, which never ceases to surprise anyone new to it, formed the point of departure for the earliest systematic investigation of refraction, which was carried out by Claudius Ptolemy, the Alexandrian astronomer and geographer, sometime during the first half of the second century A.D. Almost nothing is known about Ptolemy, and all that we have

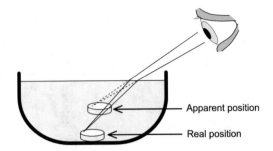

Fig. 5.5 Ptolemy's cup and coin experiment. The coin appears nearer the surface of the water because light from it refracts when it emerges into the air. The observer is not able to see this change in direction, so it appears as if light comes from a different direction than the true one

of his writings on optics is a twelfth century Latin translation of an imperfect Arabic translation from an incomplete edition of the original Greek manuscript, now lost. Fortunately, it contains an account of his optical experiments, including those on refraction.

Ptolemy was an extramissionist and so he explained the coin-in-the-cup in terms of a visual ray that issues from the eye of the observer, though his explanation works equally well in terms of intromission. He realised that the coin becomes visible when covered by water because rays of light from the coin alter direction when they pass from water to air so that they appear to reach the eye from a position above the actual coin. But by how much does refraction alter the coin's real position? This is the question that Ptolemy set out to answer. Unlike most Greek natural philosophers, he was prepared to get his hands dirty, and devised an instrument consisting of a flat, graduated bronze disc with two pivoted arms one of which was immersed in water that enabled him to measure the amount of refraction that occurs when light passes from air to water or glass.[54]

The results Ptolemy obtained with this apparatus are not far off values that can be obtained with similar apparatus today but they show signs of having been tweaked to fit a specific mathematical formula. The formula was never stated explicitly, but had it been, it might have been known as 'Ptolemy's law of refraction'. However, it would have been a law of refraction that does not accord with the now accepted law of refraction first established by Harriot and later by Descartes.

Almost a millennium later, Ptolemy's optical experiments were repeated by Alhazen. Although he too failed to discover the law of refraction, he broke new ground by attributing refraction to a change in speed. Unlike Ptolemy,

---

[54] In effect, a goniometer.

Alhazen was interested in the nature of light. He believed that light is a substance and so assumed that it must be affected by the medium through which it travels "just as a stone moves more easily and quickly in air than in water, since water resists it more than the air". In a similar manner, he argued, light must travel more slowly through glass or water than it does through air and that this is the reason light is deviated from its original path when it enters a transparent body obliquely.

Descartes also sought a physical explanation for light. In keeping with his mechanical conception of the world, he asserted that light is a pressure within the mass of tiny particles of matter that fill the spaces between the larger particles of which, he maintained, the material world is made. In other words, light does not involve the movement of matter; instead, it is like the pressure exerted by a piston compressing a liquid within a closed cylinder. Furthermore, assuming that every particle of matter is in direct contact with all its neighbours, this pressure must be transmitted instantaneously. Or, to put it another way, because there are no empty spaces in the Cartesian universe within which particles can move, something that necessarily takes time, and because Cartesian matter is incompressible, the speed of light must be infinite. Indeed, Descartes confessed to Beekman that the infinite speed of light "is so certain that if it could be proved false, I am ready to confess that I know nothing in all of philosophy".[55] What a hostage to fortune! As we shall see in Chap. 7, less than 50 years later, a Danish Astronomer, Ole Rømer (1644–1710), discovered that light takes some twenty-two minutes to travel from one side of the earth's orbit to the other, thus undermining Descartes' central tenet on the nature of matter.

When it came to explaining refraction, however, Descartes ignored both his description of light as pressure and the importance of the instantaneous propagation of light to his physics. To understand the refraction of light, he said, consider a tennis ball that has been struck by a racket at a flimsy horizontal surface (Fig. 5.6a). The ball breaks through but its speed is reduced, though only in the vertical direction—Descartes assumed that its speed parallel to the surface remains unaltered. As a consequence of the reduction in its vertical speed, the ball now travels along a path closer to the surface. In a similar manner, he suggested, light changes direction when it passes from air into glass or water. But because light refracts away from the surface, Descartes concluded that the speed of light must increase when it passes from air to water or glass (Fig. 5.6b).[56]

---

[55] Descartes, R. (1897), pp 307–8.
[56] Descartes, R. (2001), pp 77–80. English translation in: Magie, W.M. (1963), pp 265–73.

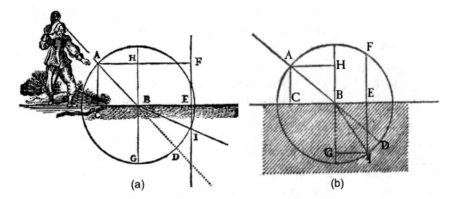

Fig. 5.6 **a** Descartes' tennis ball analogy of refraction. On encountering the surface BE, Descartes assumed that the velocity of the ball is reduced in the vertical direction and remains unaffected in the horizontal one. Hence its new direction is BI. **b** The correct geometry of the refraction of light.[57] BI is the refracted ray

Despite these questionable ideas about the nature of light and the cause of refraction, Descartes nevertheless discovered how to express the law of refraction in mathematical terms sometime between 1626 and 1628, while he was still in Paris. Details of how he found it are sketchy, due in large measure to his habitual lack of openness about his work. The only thing of which we can be certain is that he arrived at the law through mathematical analysis alone and not as the result of experiments.[58] As we now know, the law was already known to Thomas Harriot. It had also been independently discovered by a Dutchman, Willebrord van Royen Snel (1560–1626), professor of mathematics at Leiden University, a year or two before his death. Though neither man published their results, Descartes almost certainly knew of Snel's work. Posterity, however, has quite rightly given Descartes the benefit of the doubt: all the available evidence suggests that he discovered the law independently of Snel. But during his lifetime, and for several years after his death, his detractors assumed that he had plagiarised it from Snel.

Although Descartes didn't publish the law until 1637, he let Mersenne in on the secret in a letter written in June 1632:

As to my way of measuring the refraction of light, I compare the sine of the angle of incidence and the sine of the angle of refraction, but I would be happy

---

[57] Descartes, R. (2001), p 80, Fig. 10.
[58] Weinberg, S. (2015), pp 346–8.

if this were not made known yet, because the first part of my *Dioptrique* will contain nothing else but that above.[59]

It is possible that Descartes didn't want to let the cat out of the bag before he had something to show for it. His work on aspheric lenses using the law of refraction had led nowhere, though he was unwilling to admit as much. He had met with greater success when he used the same law to explain the rainbow. But to establish his reputation as a leading thinker of his day he believed that he had to show how the law of refraction followed from his ideas about matter and motion, to be published along with his essays on geometry and meteorology as a demonstration of his mathematical and scientific work.

If we take him at his word—and we have already seen that his claims should sometimes be taken with a pinch of salt—Descartes discovered the secret of the rainbow through experiments with a large spherical glass bowl filled with water.[60] He said that the inspiration for the experiment came to him while admiring a spraybow in a fountain, for it led him to realise that the cause of the rainbow is to be sought in what happens when individual drops of water are illuminated by sunlight. Fountains expressly designed to create spraybows were popular in Descartes' day and he would have had many opportunities to study the interaction between light and drops of water that creates a bow.[61] Assuming that all drops of water are spherical and that their size has no effect on the appearance of the arc, he decided that the best way to investigate the phenomenon was "to make a very big drop by filling a large glass bowl with water" (Fig. 5.7).[62]

Descartes gave few details of how he performed these experiments. He didn't say whether he held the bowl in his hands or placed it on a pedestal, though from the observations he claimed to have made it seems that he must have done both. However he set the experiment up, he found that when he turned his back to the sun and raised the bowl above his head and clear of his shadow he saw a bright red spot just inside the lower edge of the bowl when the angle between his eye, the spot of light and the direction from which sunlight entered the bowl was approximately 42°.[63] This bright spot is an image of the sun reflected from the back of the bowl and it is the source of the red band of the primary rainbow. He found that when he raised the bowl

---

[59] Descartes, R. (1897), p 255.

[60] English translation of excerpts from Descartes' account of the rainbow in: Magie, W.M. (1963), pp 273–8.

[61] Werrett, S. (2001).

[62] Descartes, R. (2001), p 332.

[63] Sunlight would have entered the bowl in the direction of the antisolar point, i.e. the middle of the shadow of his head.

Fig. 5.7 Garden spraybow. Note that the area enclosed within the bow is lighter than that outside it. The latter is the so-called dark band

ever so slightly from that position, the spot disappeared. It reappeared when he lowered the bowl, and by lowering it a little more he found that he could make the spot cycle through the other colours of the rainbow, first yellow, then blue and "other colours". At the same time he noticed a fainter spot of light near the opposite, upper edge of the bowl, which behaved much like the first one, except that it vanished when he lowered the bowl, and that it cycled through the same sequence of colours as the first spot when he raised the bowl. When he measured the angle between his eye, the fainter spot and the direction of sunlight entering the bowl, he found it to be approximately 51°. This spot is the source of the secondary rainbow.

When he moved the bowl in an arc about his head, he found that both red spots remained visible as long as the angle between his eye, the spot of light and the direction of the sun's light was 42° for the brighter spot and 51° for the fainter one. Raindrops, as he pointed out, are much, much smaller than a glass bowl, from which he concluded correctly that it's not possible to see all the colours of a rainbow in a single drop: the drops in which the red of the primary rainbow is seen are further from the centre of the bow than those in which the blue is seen and vice versa for the secondary bow. Thus it is that angles rather than distances determine the colours that are seen in each drop.

Having confirmed that the bright coloured spots seen in the bowl coincide with the position of the primary and secondary rainbows, he went on to tackle the trickier task of tracing the path of the light within the bowl responsible for the spots. He claimed that he did this by blocking the rays entering and leaving the bowl, as well as those within the water in the bowl. This, he said, is how he discovered that the bright spot is due to light that is refracted twice—once on entering the bowl and again on leaving it—and reflected once

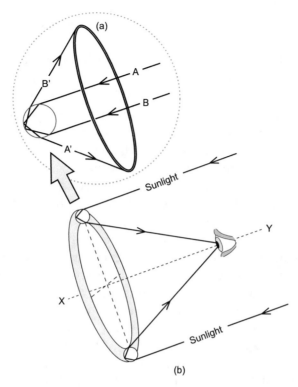

Fig. 5.8 Rainbows are circular. **a** Light reflected from within a drop emerges as a cone, the surface of which is the concentration of light responsible for the arc of the rainbow. **b** Rotating the diagram about the axis XY creates a cone. The observer's eye receives a rainbow ray from all the drops that lie on the surface of this cone. Hence the circularity of the resulting arc

from the inside surface of the bowl. As for the light responsible for the fainter spot, he found that it too is refracted twice, but that it is reflected twice inside the bowl (Fig. 5.8).

If all this sounds familiar, it is. Descartes had repeated Theodoric's experiments in their entirety, though there is no evidence to suggest that he did so knowingly. As we have seen, experiments with water-filled flasks were regularly carried out by natural philosophers interested in rainbows, some of whom may have been inspired to do so having come across snippets of Theodoric's work. As a pioneer in the science of optics, Descartes would have been well aware of these experiments before he carried out his own version, though whether his description of what he did is entirely reliable is open to question if only because, as we have seen, he had little faith in the value of experiments as a matter of principle and was, in any case, rather slapdash as an experimenter.

In the first place, he didn't believe that it is possible to discover anything new through experiments. At best, he maintained, an experiment merely confirms what is already known, so if an experiment produces results that don't agree with a theory, the experiment is at fault, not the theory. Experimental results could never challenge conclusions derived from "clear and distinct ideas", though they could support them. So, although the experiments with the bowl undoubtedly proved useful to him in confirming some of facts about rainbows, it is most unlikely that he determined the path of the rays responsible for the rainbow experimentally.

There are also practical reasons for doubting Descartes account of how he determined the path of the rays responsible for the bright spots. While it's possible to block the light entering and leaving the flask—it can be done by placing a finger at the right place on the surface of flask—it is not possible to block the beam within the glass bowl with the degree of precision implied in his account.[64] At best, the procedure he describes would have enabled him to determine that the light responsible for the rainbow enters a drop very near one edge and leaves from the opposite edge. Given the difficulty of performing this part of the experiment, it seems more than likely that Descartes determined the path of the rays of light by calculation. Why he thought that his explanation would carry more weight if it appeared that the path of the light through the drop was first determined experimentally remains a mystery, especially given his reservations about the value of experimental investigations.

The experiment with the bowl may have confirmed important facts about the rainbow, but as Descartes realised, it left their explanation open.

> But the principal difficulty still remained, which was to understand why, since there were many other rays there which after two refractions and one or two reflections, can tend towards the eye when the globe is in another position, it is nonetheless only those of which I have spoken that cause certain colours to appear.[65]

Why indeed? Descartes had put his finger on the problem. But his commitment to a mechanical explanation of the world meant that if he was to find

---

[64] It is, however, possible to prevent the internal ray being reflected by pressing the tip of a wet finger on the outside surface of the flask at the spot where the reflection occurs, thus identifying where that takes place. This is, nevertheless, a rough and ready method, and would not allow one to determine the path of a ray as accurately as Descartes claims to have done. For experimental details, see: Minnaert, M. (1954), p 175.

[65] Descartes, R. (2001), p 334.

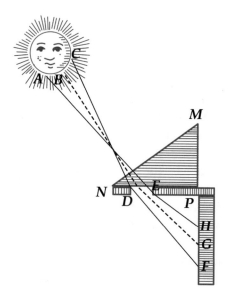

Fig. 5.9 Descartes' prism experiment. Descartes was the first person to publish a description of an experiment in which colours are produced by refracting light through a prism[67]

the answer, he would also have to find an alternative to Aristotle's ideas about colour, ideas that were widely accepted at the time.

Descartes pointed out that the colours in a rainbow are also seen in sunlight that has passed through a glass prism. He described an experiment in which he had arranged a right-angled prism so that sunlight fell perpendicularly on the hypotenuse (the longest face of prism). In this way he ensured that the light emerging from the opposite face of the prism was refracted only once. It also eliminated reflection as a cause of the rainbow's colours because light that emerges from the prism in this situation has been not reflected. To produce a spectrum (a word he did not use, it was coined by Newton), Descartes found that he had to cover the lower face of the prism with a card in which there is a small hole: without this, no colours were visible—simply a spot of bright, uncoloured sunlight (Fig. 5.9).[66]

Having performed an experiment that is strikingly similar to the one famously carried out by Newton some thirty years later, Descartes misinterpreted the results. Where Newton concluded that the colours in the spectrum are already present in white light and that they are separated from one another as they pass through a prism because each colour is refracted to a different

---

[66] Descartes, R. (2001), p 335.

[67] Descartes, R. (2001), p 335, Fig. 20.

degree, Descartes was thrown off the scent because he found that when the hole in the card was made larger, the colours at the centre of the spectrum, yellow and green, retreated to opposite edges, leaving a bright colourless spot fringed on one side with red and on the other with blue. He attributed these colours to the presence of the edge, or shadow as he called it, of the hole through which the light passed when it emerged from the prism.

If only he had resisted his penchant for theorising, and pressed on with the experiment, he might have realised that the most significant fact is that each colour differs from its neighbour in the degree to which it is refracted by the prism, and pipped Newton to the post. But given Descartes' cavalier attitude to of the value of experiment, it was perhaps inevitable that he was content to base his theory about origin of colour on a single experiment, whereas Newton performed dozens of experiments, each one of which followed from a previous observation and which was designed to explore its implications, before he felt sufficiently confident to draw any conclusions about the origin of the colours.

So instead of performing further experiments to discover why a spectrum is produced when sunlight is refracted by a prism, Descartes resorted to an explanation based on his analogy between light and tennis balls. He assumed that white light consists of particles whose "turning motion is nearly equal to their motion in a straight line."[68] He then tried to show how the rotation of these particles is either increased or retarded when they strike the edge of the card on leaving the prism. He concluded that colour depends on how fast they rotate: the fastest rotation gives rise to red and the slowest to blue. Or in the language of mechanical philosophy, the sensation of colour is due to the effect of the rate of spin of matter on the nervous system responsible for vision.

Descartes' explanation for the origin colour may have been wrong in every respect, but it had one thing in its favour: it broke with the Aristotelian distinction between real and apparent colours. If the colours in a rainbow can be explained mechanically then they are no different to the colours of solid objects. However, like the Aristotelians, Descartes assumed that colours are brought about through the modification of white light. As we saw in the last chapter, according to Aristotle the colours in the rainbow are brought about when white light is modified by being weakened. For Descartes the modification takes place when light encounters an edge that alters the rotation of luminous matter. Nevertheless, in both accounts, colours are the result of a bruising encounter between unblemished white light and gross matter.

---

[68] Descartes, R. (2001), p 336.

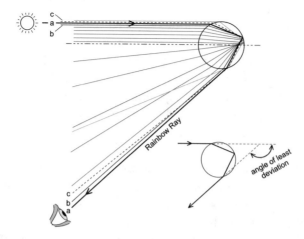

Fig. 5.10 Refraction of several parallel rays of light through a spherical drop. The rainbow ray, *a*, is the ray that is deviated least from the direction in which all rays enter the drop. Every other ray undergoes a greater deviation. Notice that a ray such as *b* emerges from the drop almost parallel to the rainbow ray, so lies within the bundle of rays that contribute to the caustic which is the source of the narrow arc of the rainbow. Rays such as *c* that enter the drop above the rainbow ray are deviated more than those below such as *b*

We now come to Descartes' *pièce de la résistance*, upon which rests his legitimate claim to have explained the rainbow. It begins unpromisingly. Where, he asked himself, is the edge or shadow within a drop necessary to bring about the rotation responsible for the rainbow's colours? But he followed this up immediately with a far more fruitful question, one that required a mathematical answer: "why [colours] appeared only under certain angles."[69]

Using the laws of reflection and refraction, he laboriously calculated the path of several equally spaced parallel rays through a notional drop of water. He found that as the angle at which a ray meets the surface of drop increases, the angle at which it emerges from the drop in the direction of the observer, measured from the direction from which the sun's rays enter the drop at first increases, reaches a maximum value of approximately 42° and then decreases. The ray that emerges at the largest angle marks the theoretical outer edge of the primary rainbow.[70] It is sometimes referred to as the rainbow ray despite being no more than a useful artefact of ray-tracing (Fig. 5.10). Being merely a line, however, the rainbow ray can't account for the rainbow's arc.

---

[69] Descartes, R. (2001), p 339.

[70] Confusingly, the angle at with the rainbow ray emerges from a drop is more correctly known as the angle of *minimum* or *least deviation* because it is measured from the direction of the rays that enter the drop.

More significantly, his calculations revealed that there is a bundle of rays, amounting to some 10% of the light reflected within a drop, that emerge between the angles of 41 and 42°. This concentration of rays, the technical term for which is a caustic surface, is responsible for the bright narrow arc of the primary bow.[71] All other rays emerge from the drop within the arc of the primary bow, which is why the sky enclosed by the bow is sometimes noticeably brighter that that outside the bow.

When he repeated the calculations for rays that were reflected twice within the drop, he found that there is another caustic surface between 52 and 54°, which corresponds to the secondary bow. Furthermore, no doubly reflected rays emerge at angles less than 52°. The absence of reflected light between the outer edge of the primary bow and the inner edge of the secondary bow is the reason why the sky between the bows appears darker than that enclosed by the primary bow.

So here, at long last, almost 2000 years to the day after Aristotle's death, with the exception of Thomas Harriot's unpublished account, was the first satisfactory mathematical explanation for the size and shape of a rainbow, if not for its colours. Whereas most of the light reflected within a drop spreads out as a broad cone, the primary arc of a rainbow is due to a concentration of light between 41 and 42°. The outer edge of this arc is formed by light that is deviated least after one reflection from the direction it enters a drop. Or, to put it another way, outer edge of the primary bow stands out clearly because no light emerges from a spherical drop after one reflection from the side of the drop facing the observer at an angle greater than 42°.[72] The absence of light beyond this point is the explanation for Alexander's dark band.

Descartes suggested that the dark band is the cause the bow's spectral colours. He assumed that the absence of light beyond the arc provides the edge or shadow necessary to make sunlight red. The blue edge of the arc is due to the shadow that lies beyond the rays that emerge at 41°, a shadow that is due to a reduction in light intensity at angles less than 41°.

As we have seen, although he had observed colours fanning out from a prism, Descartes never realised that there might be a link between colour and the degree of refraction. Consequently, he based all his calculations on a single value for the refractive index of light when it passes from water to air. He took this to be 187 to 250 (i.e., 0.75), which, post-Newton, we know corresponds to the refractive index of green light.[73] By using a single value for refractive index, the most Descartes could achieve was to show that there

---

[71] The word *caustic* comes from the Greek for "burnt". See the appendix for more on caustics.

[72] We shall see in Chap. 8 that this is not entirely correct.

[73] Descartes, R. (2001), p 340.

is a concentration of rays between 41 and 42° and therefore explain why the arc of a rainbow is a narrow band rather than a pencil thin line.

It's obvious from his account that he didn't check his theoretical result against a natural rainbow. Had he done so he would have realised that something was amiss. What he had done is to calculate the maximum angular radius of the primary bow, and the minimum angular radius of the secondary bow. He assumed that the concentration of light between 41 and 42° accounted for the entire width of the bow, whereas, had he measured the width of a natural rainbow, he would have realised that the primary arc is in fact more than twice as wide, i.e. 2° wide, about the apparent width of a finger seen at arms-length.

Of course, Descartes would have been stumped at this point, since it never occurred to him that refraction is responsible for the rainbow's colours. He never realised that the blue band in a rainbow is due to a caustic surface of blue light between 40 and 41°. Perhaps he did measure a bow and, finding that it was wider than his calculations indicated, threw in the towel because he had no means of explaining this, without admitting as much to the world at large, something that would have gone against the grain of someone with his disposition. Once again, his belief that theory trumps experience let him down.

Descartes concluded his account of the rainbow by proclaiming that he had solved the riddle of the rainbow and that it only remained to tidy up some loose ends.

> Thus, I believe that no difficulty remains in this matter, unless it perhaps concerns the irregularities which are encountered; for example, when the arc is not exactly round or when its centre is not in the straight line passing through the eye and the sun, which can happen if the winds change the shape of the raindrops; for they may not lose the smallest part of their roundness without this making a notable difference in the angle under which the colours must appear.[74]

He was quite right: the shape of a raindrop has a noticeable effect on the appearance of a rainbow, but this was not something that was comprehensively investigated until the twentieth century, and which we shall consider in greater detail in Chap. 7.

Surprisingly, Descartes' work on the rainbow had a limited impact. Several accounts of rainbows written in the decades following publication of the *Discourse* either ignored his explanation or attacked it. To add insult to injury,

---

[74] Descartes, R. (2001), p 343.

in old age, Isaac Newton, who knew better, maliciously attributed the major advances in explaining the rainbow since Aristotle to Dominis of all people, and said that Descartes had merely built on Dominis' discoveries.

One of the few natural philosophers of note to grasp the significance of Descartes' explanation of the rainbow was Christiaan Huygens, the doyen of European science of his day. Almost 50 years after the publication of the *Discourse*, in a review of the first biography of Descartes,[75] Huygens wrote

> The prettiest thing which [Descartes] found in physical matters, and in which perhaps his view is well taken, is the cause of the double arc of the rainbow - i.e. the determination of their angles and apparent diameters.[76]

Huygens went on to say that the efforts of Descartes' contemporaries on the subject of the rainbow were "pitiful", but dismissed Descartes' explanation of its colours out of hand as "nothing less probable."

In fact, Descartes made several original contributions to our understanding of the rainbow. As Huygens recognised, the most important of these was to have determined precisely the paths of rays of light through a spherical drop of water, thus explaining the size, shape and position relative to the observer of both the primary and secondary arc.

Descartes should also be given credit for the diagram that accompanies the explanation, showing the large spherical vase he used superimposed on a curtain of rain. It is an eye-catching visual explanation of a rainbow, showing as it does the path of rays responsible for both the primary and secondary arcs within an idealised magnified drop, the position of the antisolar point and position of the arcs relative to the observer. Although Descartes didn't draw the diagram—he acknowledged that he was a poor draughtsman—he instructed and supervised the engraver who did; it is evidence of Descartes' powerful visual imagination and has been the template for scientific diagrams of the rainbow ever since (Fig. 5.11).

On the other hand, his law of refraction, the cornerstone of his explanation of the rainbow, was not based on any insights into the nature of light, despite his claims to the contrary. It doesn't tell us why refraction occurs, so its use to explain the features of the rainbow isn't a huge step forward from what can be established through careful measurement of the angles of incidence and refraction for several individual rays. As far as the explanation of the rainbow is concerned, his law of refraction was no more than a calculating aid. The same results can be achieved by carefully measuring corresponding

---

[75] Baillet, A. (1691).
[76] Huygens, C. (1905), p 405.

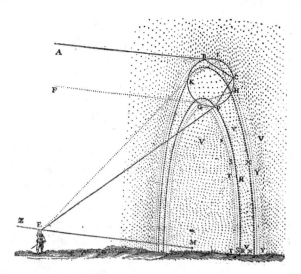

Fig. 5.11 Descartes' rainbow diagram. It shows the path of rays through a larger than life raindrop at the apex of the arc of a rainbow. The observer is at E. ABCDE is the rainbow ray responsible for the primary bow. FGHIKE is the rainbow ray responsible for the secondary bow[77]

angles of incidence and refraction in an experiment in which a succession of narrow beams of light pass from air to water and the measurements plotted as a graph. The path of any ray through a spherical drop can then be determined with reasonable accuracy by reading off the graph the angle of refraction that corresponds to the angle of incidence which the ray makes as it enters the drop.

But, of course, this would be an immensely tedious procedure. Descartes' task was eased enormously by the law of refraction because it enabled him to calculate angles of refraction for very small intervals of the angle of incidence, thus determining the path of as many rays as he required. He found that he had to perform more than a hundred and fifty separate calculations to establish that a rainbow is due to a bundle of rays all of which emerge from a drop at almost the same angle to create a caustic surface bright enough to be visible against a bright background such as clouds.

Indeed, far from the law of refraction providing the explanation for the rainbow, the reverse is the case: the explanation of the rainbow helped confirm the law of refraction. Descartes himself cheerfully admitted that he

---

[77] Wikimedia Commons public domain: https://tinyurl.com/2p89tv23 (accessed 8/10/2022).

had never carried out a direct experimental test of the law.[78] The only circumstances in which he had used the law of refraction before he came to explain the rainbow in 1629 was to determine the theoretical curvature of an aspheric lens, probably in 1628.

Descartes claimed on several occasions that such a lens had been made to his design and that it performed as expected, which, he said, vindicated his law of refraction. But his claim has never been corroborated. Most studies of the development of lenses and telescopes have concluded that no such lens was made, or could have been made, during the seventeenth century, which leaves the explanation of the rainbow as the only concrete proof that Descartes had to offer for the truth of law of refraction, even though he never considered it to be so.[79]

Given his hopelessly muddled ideas about the nature of colour, Descartes had taken the explanation of the rainbow as far as he was able. Someone else would have to solve the problem of its colours, even though at the time this was not seen as an issue that had to be addressed. That person was Isaac Newton.

Descartes' explanation of the rainbow has a further claim to fame. It was, arguably, the first published example of a successful application of mathematical physics to a natural phenomenon. Credit for the earliest successful application of mathematical physics is quite rightly given to Galileo for showing that the trajectory of a body projected into space near the earth's surface should be a parabola. Galileo had indeed solved the problem of parabolic motion more than two decades before Descartes had found the explanation of the rainbow, but publication of Galileo's book was delayed until 1638 due to his problems with the Inquisition, whereas Descartes' *Discourse* was published a year earlier in 1637. Tellingly, both books were published in the same town, Leiden in Holland, which was well beyond the reach of the Inquisition.[80]

In both cases, Descartes' explanation of the rainbow and Galileo's explanation of parabolic motion, a natural phenomenon was explained by employing general laws of nature established independently of the phenomenon they were used to explain and expressed in mathematical form: Descartes' law of refraction and Galileo's law of uniformly accelerated motion. Unlike Galileo's

---

[78] Letter to Jakob Golius, 2 Feb, 1632. In: Descartes, R. (1897), pp 236–42. Golius (or Gool) was professor of mathematics at Leiden University, where Descartes enrolled as a student when he first went to live in Holland. Golius was Willibrord Snel's successor in the chair and it was he who found out that Snel had already discovered the law of refraction. But Golius never suggested that Descartes had plagiarised Snel.

[79] Helden, A.v. (1974).

[80] Galilei, G. (1954), pp 244–95.

proof that a body projected through space follows a parabolic path, however, Descartes' account of the rainbow yields results that agree closely with what we see in nature whereas, due to the effect of air friction, the path of a body travelling freely through the atmosphere is never perfectly parabolic.

Descartes was full of praise for Galileo's explanation of parabolic motion but maintained that Galileo had "built without foundation." In Descartes' view, the central aim of a natural philosopher should be to go beyond mathematical descriptions of phenomena and explain their physical causes. Galileo, said Descartes, had merely provided mathematical descriptions of motion while ignoring their physical causes. Nevertheless, Galileo's experimental and mathematical approach to nature, free from underlying metaphysical considerations, proved to be far more fruitful than Descartes supposed, and provided a template for the major scientific advances of the seventeenth century, particularly Newton's work on motion, force and gravity.

Descartes' influence was less direct, though equally significant. His mechanical philosophy, the idea that the entire universe, organic and inorganic, can be explained in terms of the direct action of one bit of matter on another, did away with occult forces and arbitrary Aristotelian distinctions such as that between the celestial and sublunary spheres or between real and apparent colours. Where Galileo was prepared to accept that the laws that governed the heavens were different from those the applied to terrestrial phenomena, Descartes advocated the unity of nature, in which the only reality was matter and motion. Hence every event could, in principle, be explained in the same simple terms: the action of one bit of matter on another. Descartes' lasting legacy was the idea that there are no unfathomable mysteries in the physical world because a wholly mechanical account of cause and effect of any event in nature is possible in principle.

Unfortunately, he took this to mean that principal task of science is to come up with plausible mechanisms to explain particular phenomena, rather than the investigation of nature through a combination of experiment and mathematics. As we have noted, with few exceptions, in his hands this led to all sorts of improbable explanations that often flew in the face of common sense, let alone experimental evidence.

The way forward was to abandon the search for fanciful mechanical explanations based on plausible assumptions and rely instead on mathematics and experiment in the spirit of Galileo.

6

The Celebrated Phenomena of Colours

> To perform my late promise to you, I shall without further ceremony acquaint you, that in the beginning of the Year 1666 ... I procured me a Triangular glass-Prisme, to try therewith the celebrated Phænomena of Colours.[1]

One way or another, Descartes' mechanical philosophy influenced almost every major seventeenth century natural philosopher including the man destined to be his nemesis, Isaac Newton. Newton, the name, if not the man himself and his works, needs no introduction as the epitome of scientific genius. But, disconcertingly for the legions of his secular admirers, Newton is not so easily pigeonholed, for he employed his formidable intellect chiefly as a theologian and alchemist; science and mathematics were sideshows to him. Indeed, had he been asked to state his vocation, it is likely that he would have replied unhesitatingly "scholar"—not "mathematician" or "natural philosopher". Yet such was the reach of his powerful mind, that during the two or three years following his graduation from Cambridge University in 1665, several years before he became consumed by research into alchemy and theology, he single-handedly laid the foundations of modern physics and mathematics, "For in those days I was in the prime of my age of invention and minded Mathematics and Philosophy more than at any time since."[2]

---

[1] Newton, I. (1671/2 b), p 3075.
[2] Westfall, R.S. (1980), p 143.

© The Author(s), under exclusive license to Springer Nature Switzerland AG 2023
J. Naylor, *The Riddle of the Rainbow*, Copernicus Books,
https://doi.org/10.1007/978-3-031-23908-3_6

Fig. 6.1 Rainbow over Newton's house in Woolsthorpe[4]

Born on Christmas day, 1642,[3] in Woolsthorpe, a village near Grantham in Lincolnshire, he was the only son of an illiterate farmer who died a month before his son was born. Isaac showed signs of an inquisitive mind and dextrous hands at a young age. He attended Grantham Grammar School where he was remembered as 'always a sober, silent, thinking lad' and for his genius for constructing mechanical devices. But his mother's ambitions for him went no further than that he should take over the family farm as soon as he was old enough. When he turned seventeen, she took him out of school and put in charge of the farm for a few months, a role for which he proved spectacularly unfit. Reluctantly, she agreed with her brother, a clergyman, that Isaac would be better off at university (Fig. 6.1).

Cambridge University, where Newton enrolled in 1661 when he was eighteen, was not the intellectual powerhouse it was to become in the nineteenth century.[5] Lecturers neglected their duties and students were left to fend for themselves. One complained "I had none to direct me, what books to read,

---

[3] At the time of Newton's birth, England was still using the Julian calendar, which was then 10 days out of step with the modern Gregorian calendar that had been adopted by Catholic countries several decades earlier. Newton's 'real' birth date was thus 4 January, 1643. Protestant England refused to fall into line with the Continent because it believed that the Gregorian calendar was a Popish plot. To further complicate matters, the English new year fell on 25 March. Hence the English often gave two years for months falling between January and April: e.g. Jan 1642/3. England eventually adopted the Gregorian calendar on 14th September 1752, by which time the discrepancy amounted to 12 days. The first country to adopt the Gregorian was Spain on 15 October 1582, Russia the most recent (31st January 1918).

[4] Photo by Roy Bishop, used with permission.

[5] Newton was a student at Trinity College.

or what to seek, or in which method to proceed."[6] Just the sort of place to send a budding genius. It was to suit Newton down to the ground because it left him to his own devices and gave him the freedom to follow his own interests.

Even by the standards of the time, the University's curriculum was moribund, if not obsolete. It was based largely on Aristotle, consisting of logic, ethics and rhetoric, but little science and even less mathematics. It may have suited those destined for the law or the church, but provided slim pickings for anyone of a scientific or mathematical bent. Newton's student notes show that he never finished reading any of the prescribed texts, something that was to cost him a good degree. The notes also reveal the genesis of his genius. By 1664 he had all but abandoned the university syllabus and struck out on his own, for a genius, perforce, must be an autodidact. Among other things he became intensely interested in optics and read Robert Boyle's "Experiments And Considerations Touching Colours".[7] One of these experiments in particular struck a chord: it involved refracting a narrow beam of sunlight through a prism within a darkened room. In Boyle's opinion, a glass prism was "...the usefullest Instrument Men have yet imploy'd about the Contemplation of Colours", a claim that may well have been the inspiration for Newton's own experiments with prisms.[8]

Another book that made a profound impression on Newton was Robert Hooke's 1665 *Micrographia, or Some Physiological Descriptions of Minute Bodies Made by Magnifying Glasses,*[9] an eclectic cornucopia of observations and ruminations, which, as the subtitle intimates, is largely devoted to descriptions of Hooke's microscopic studies of insects, plants and small inanimate objects, and illustrated with wonderfully detailed drawings—Hooke was a meticulous observer and a first-rate draughtsman, and several of the drawings in *Micrographia* are well-nigh photographic in their detail. Newton was particularly intrigued by Hooke's investigation of colours seen in thin sheets of Muscovy Glass, a transparent mineral now called mica, in soap bubbles and in thin films of air or water trapped between two sheets of glass. Hooke claimed that Descartes' theory of light and colour could not explain these colours and came up with an alternative: light is a "vibrative motion" within a transparent medium, like the ripples created when a stone is dropped into still water, and not a stream of rotating Cartesian globules.[10]

---

[6] Hall, A.R. (1992), p 17.
[7] Robert Boyle (1627–91) was a natural philosopher with eclectic interests.
[8] Boyle, R. (1664), pp 228–9.
[9] Hooke, R. (2003).
[10] Hooke, R. (2003), pp 47–67.

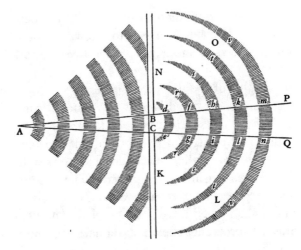

Fig. 6.2 Newton's illustration of diffraction of waves in water traveling from A to PQ, after passing through a narrow opening, BC[13]

Newton was not persuaded. Lights cast sharp shadows but sounds can be heard around a corner (Fig. 6.2). If light is a vibration, he wrote in his notebook, "Why then may not light deflect from straight lines as well as sounds &c?".[11] It is a matter of observation that vibrations in a medium such as ripples in water wrap around the edges of an object: "That these things are so, anyone may find by making the experiment in still water ... And we find the same by experience also in sounds which are heard through a mountain interposed; and, if they come into a chamber through the window, dilate themselves into all the parts of the room, and are heard in every corner; and not as reflected from the opposite walls, but directly propagated from the window, as far as our sense can judge."[12] Newton was right about sound but drew the wrong conclusion concerning light. To the end of his life, he remained convinced that it isn't possible to explain the fact that light travels in straight lines if it is assumed that it is a vibration.

But the person from whom Newton learned most was Descartes. Descartes' *Principia Philosophiae* introduced him to Cartesian metaphysics, cosmology and mechanics. As the title suggests, this work was written in Latin, which meant that Newton was able to read it. It was published in 1644, the same year that saw Latin translations of Descartes' *Optics* and *Meteorology* from which Newton learned the law of refraction and its application to the design

---

[11] Hall, A.R., Boas Hall, M. (1962), p 403.

[12] Newton, I. (1960), Prop 42, Theorem 33, Case 2.

[13] Newton, I., (1960) p 369.

of lenses and the explanation of the rainbow. And the 1649 Latin edition of Descartes's Geometry introduced him to the ideas and methods of European mathematicians.

These extra-curricular interests threatened Newton's academic prospects. Rather late in the day, he realised that if he was to continue with his own research into mathematics and optics he had first to secure his future at the university. He returned to the prescribed texts in a last-minute preparation to qualify for a scholarship that would enable him to stay on at Cambridge for a further four years. He was awarded a scholarship in April 1664, if only by the skin of his teeth, having failed to impress his examiner, Isaac Barrow (1630–1677), with his knowledge of mathematics. Barrow was the university's first professor of mathematics and the Chair he occupied had been endowed in 1663 by Henry Lucas, hence its holder is known as the Lucasian Professor. Barrow had examined Newton on Euclid's geometry, which Newton, having spent his energies mastering Descartes' analytical geometry, had apparently neglected to study in detail.

If only Barrow had asked Newton to expound on Cartesian geometry he might have got the measure of the man there and then because Newton was first and foremost a mathematician. Accordingly, the earliest flowering of his genius was in mathematics. In the twenty months between his election to a scholarship in April 1664 and January 1666 he came up with the calculus, a mathematical method for handling changing quantities, which was to be the key to his later work on forces and gravity and, indeed, to his explanation of the rainbow. Thus, within a year of his graduation in January 1665, having just turned twenty-three, Newton had already outstripped every other mathematician of his generation. But by not publishing anything on the subject at the time, the world at large was unaware of either his talent or his discovery.

Such genius came at a price. Newton was single-minded to the point of obsession. When asked in later life how he discovered the law of gravitation he is said to have replied, "By thinking about it continuously." Unsurprisingly, his contemporaries found him taciturn, reclusive, and secretive. And while such traits might be put down to the exigencies of genius, he displayed other character flaws that were altogether all too human. He was quick to take offence, insecure and unforgiving; "…a little too apt to raise in himself suspicions where there is no ground", according to John Locke, the philosopher, who knew Newton well in his latter years.[14] Not someone whose company one would seek out, unless it was to discuss natural philosophy or mathematics. Accordingly, few outside a tiny circle in Cambridge were aware of

---

[14] Locke, J. (2007). Letter to Peter King, 30th April 1703.

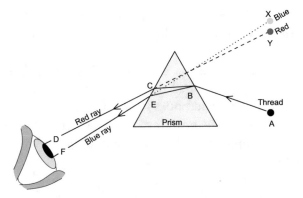

Fig. 6.3 Newton's experiment with coloured threads. Refraction at B causes 'red rays' and 'blue rays' to travel in slightly different directions through the prism. Hence 'red rays' emerge at C, enter the eye at D and appear to come from Y. The 'blue rays' emerge at E, enter the eye at F and appear to come from X

his gifts or of his achievements. Fortunately, among those who had come to recognise the magnitude of Newton's mathematical genius was someone who was in a position to further his career: the Lucasian Professor, Isaac Barrow.

Newton's earliest optical experiments date from this period and involved looking at things through a glass prism. Early in 1665 he noticed, as had others before him, that seen through a prism everything is fringed with colour, violet and blue along one edge and red and yellow along the other. This observation inspired the first of several optical experiments. He tied together a blue and a red thread and stretched the combination out on a sheet of black paper. When he looked at the threads through a prism he saw the blue thread displaced towards the apex of the prism and the red thread displaced towards its base so that the threads no longer appeared to be joined at the knot. Yet the red thread remained red and the blue one remained blue. Far from the prism modifying white light to produce colours as Descartes and others maintained, Newton concluded that colours are already present in white light and that a prism merely separates these colours from one another because it refracts each one to a different degree to all the others. Consequently, light from the blue thread emerges from the prism in a slightly different direction to light from the red thread (Fig. 6.3).

1665 was also the year of the Great Plague, 'the death' as it was then called, which had first taken hold in London. By July 'the death' had reached Cambridge and the Colleges were closed; those who could departed for the safety of country. Newton took refuge in his family home in Woolsthorpe, taking with him his books and notes so that he could continue his researches. Despite his later claims, however, it seems that he never carried out any optical experiments while he was at Woolsthorpe, concentrating instead on developing his 'method of fluxions', i.e. the calculus, and, according to his own account, inspired by the sight of an apple falling from a tree, to begin musing about the nature of gravity. He remained in Woolsthorpe until early 1667.

He recalled his prism experiments the following year while attempting unsuccessfully to grind a lens free from the defect that Descartes had sought to eliminate: the inability of a spherical lens to focus all parallel rays to the same point. As he worked on his lens, it must have dawned on Newton that he was on a hiding to nothing. Looking through a lens is rather like looking through a prism because, like those of a prism, the opposite faces of a lens are inclined to one another. Even if he could grind a lens with the shape specified by Descartes, such a lens would still not focus all rays to a single point because each colour is refracted by the glass to a different degree from all the others. This is why the images of stars and planets seen through the refracting telescopes of the time were always fringed with colours, making it impossible to see them clearly. The only way to eliminate these coloured fringes in telescopes, Newton concluded, was to abandon lenses altogether and use mirrors because, unlike refraction, reflection does not bring about a separation of colours as long as the reflecting surface is perfectly smooth. And so he turned to designing and constructing the world's first reflecting telescope, a task he completed in August 1668. It was tiny—the tube a mere 15 cm long and 2.5 cm wide—but, he claimed, it magnified "about 40 times in Diameter which is more than any 6 foote Tube can doe."[15]

A year later, in October 1669, on Barrow's recommendation, Newton was appointed Lucasian Professor in Barrow's place, a position he was to hold for the next thirty years. Barrow declaring that he wished "to serve God and the Gospel of his son", went on to better things; within a year of giving up his professorship he was appointed chaplain to Charles II, and three years later was made master of Trinity College, Cambridge.

The holder of the Lucasian Professorship was required to lecture once a week during term time on "some part of Geometry, Astronomy, Geography, Optics, Statics or some other Mathematical discipline." Drawing on the research he had already carried out into light and colour, the topic Newton chose was optics.[16] In these lectures he set out his ground breaking ideas on light and colour, including an explanation for the colours of the rainbow. Inevitably, given the laisser-faire attitude of Cambridge students of the time,

---

[15] Newton, I. (1668/9). In: Turnbull, W.H. (1959), pp 3–4.

[16] Barrow's last series of university lectures was on optics, and Newton helped him revise them for publication.

his lectures were poorly attended: "so few went to hear him, and fewer that understood him, that oftimes he did, in a manner, for want of Hearers, read to the Walls."[17] In fact, even if the lecture hall had been full to capacity with eager listeners, Newton's lectures would have had limited impact because few undergraduates of the time would have had the wherewithal to understand what he was saying, let alone follow the way in which he said it. Not only was he setting forth radically new ideas about light and colour, his exposition was based on a combination of experiments and mathematics, something that would have been unfamiliar to an audience schooled in the methods of Aristotelian natural philosophy.

A few months after his appointment to the Lucasian Professorship, Newton made a second reflecting telescope, an improved and slightly larger version of the first one. Colleagues at the University and others urged him to show it to the Royal Society, if only to claim priority and protect his invention "from yᵉ Usurpation of forreiners".[18] Barrow took it to London where it caused a sensation; even the King, Charles II, took an interest. Newton was immediately elected a Fellow of the Royal Society on the strength of it. The enthusiastic reception of the instrument prompted this reticent and private man to do the unthinkable: he volunteered to provide the members of the Royal Society with an account of the "Philosophical discovery which induced me to the making of said telescope."[19] The offer was eagerly taken up, and Newton's account of his "philosophical discovery" was read on his behalf to a meeting of the Society on the evening of Wednesday, 8 February, 1672. He himself remained in Cambridge, doubtless anticipating a favourable reception of his ideas on the nature of light and colour that would cap the triumph of his telescope.

Newton called his account "A New Theory About Light and Colours". It was certainly new, for the discovery he wanted to share with his peers was that white light is an amalgam of several distinct colours, something that even today, when we are familiar with Newton's ideas, seems to fly in the face of common sense. How can a combination of several vivid, coloured lights create something that is colourless, i.e. white? As we shall see in Chap. 10, the answer to this question wasn't found until long after Newton's death.

He began with a brief description of the observation that had led to his discovery.[20]

---

[17] Newton, H. (1727/8), p 1.

[18] Oldenburg, H. (1671/2). In: Turnbull, W.H. (1959), p 73.

[19] Newton, I. (1671/2 a). In: Turnbull, W.H. (1959), p 82.

[20] The exact place, time and circumstances of this experiment has never been satisfactorily determined. Westfall points out that it could not have taken place "at the beginning of the year" because the

To perform my late promise to you, I shall without further ceremony acquaint you, that in the beginning of the Year 1666 (at which time I applied my self to the grinding of Optick glasses of other figures than Spherical,) I procured me a Triangular glass-Prisme, to try therewith the celebrated Phænomena of Colours. And in order thereto having darkened my chamber, and made a small hole in my window-shuts, to let in a convenient quantity of the Suns light, I placed my Prisme at his entrance, that it might be thereby refracted to the opposite wall. It was at first a very pleasing divertisement, to view the vivid and intense colours produced thereby; but after a while applying my self to consider them more circumspectly, I became surprised to see them in an oblong form; which, according to the received laws of Refraction, I expected should have been circular.[21]

The idea of using a prism to investigate the nature of light was not new. As we saw in the last chapter, Descartes had based his explanation of the cause of colour on a similar experiment. And Boyle had used a prism in a darkened room to create what he called a 'Prismatical Iris'. But both Descartes and Boyle had tacitly assumed that light, whatever its colour, is always refracted to the same degree, which, as Newton knew from his optical researches, entails that the sun's refracted image should be circular when a prism is arranged so that sunlight passes through it symmetrically. Yet he could clearly see that its image is, in fact, drawn out into an oblong that is several times longer than it is broad (Fig. 6.4).

Newton realised that this elongation was highly significant and set about discovering its cause. He refracted a narrow beam of sunlight through different parts of a prism to see if the thickness of the glass had any effect on the elongation; it didn't. Nor did irregularities and imperfections within the glass. What of the fact that light from one edge of the sun's disc enters the prism from a slightly different direction than light from its opposite edge? This, too, was eliminated as a cause of the elongation simply by rotating a prism about its horizontal axis to vary the angle at which the beam of sunlight entered the prism. Newton found that he could rotate a prism through several degrees with very little effect on the length of the spectrum, thus ruling out the small difference due to the sun's angular size as the cause of the elonga-tion. He even put Descartes' ideas on the cause of colour to the test. Newton had noticed that the trajectory of a tennis ball given a spin, curves slightly. He reasoned that the faster spinning globules that Descartes claimed were responsible for redness might be expected to follow a path having a slightly

---

maximum height reached by the Sun is about 20°, which is not high enough to allow Newton to cast a spectrum on a distant wall. See: Westfall, R.S. (1980), pp 156–8.
[21] Newton, I. (1671/2 b), p 3075.

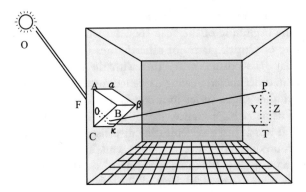

Fig. 6.4 "In a very darkened Chamber, at a round Hole, about one third Part of an Inch Broad, made in the Shut of a Window, I placed a Glass Prism, whereby the Beam of the Sun's Light, which came in at that Hole, might be refracted upwards towards the opposite Wall of the Chamber, and there form a colour'd Image of the Sun."[22]

greater curvature than slower spinning ones. This would result in an elongated spectrum, yet careful measurement established that whatever its colour, light follows a straight path after leaving the prism and so Descartes must be mistaken in attributing colour to the rotation of his globules.

But in Newton's opinion one of his optical experiments stood out from all the others, the *experimentum crucis* as he called it, brazenly appropriating a phrase coined by Hooke in *Micrographia*. Hooke's *experimentum crucis* was the presence of colours in thin films, which he believed could be used "to direct our course in the search after the true cause of Colours."[23] Newton also sought the true cause of colours, but believed that Hooke had settled on the wrong experiment. In fact, as we shall see in the next chapter, Hooke's *experimentum crucis* revealed far more about the nature of light and colour than Newton's, but that was something that didn't become apparent for almost another century and a half.

In a darkened room, wrote Newton, he had drilled a ¼ inch (6 mm) circular hole in one of the window shutters and placed an inverted prism against the hole, adjusting its position so that it cast a coloured spectrum of the sun on the wall some 23 feet (7 m) away at the far side of the room. By placing a board with a small hole in the path of the spectrum he was able to select a narrow beam of a single colour that could be refracted by a second prism on the other side of the board. If refraction modified light to create colour, as Descartes and others had claimed, the second prism should noticeably alter the colour of the selected beam. But the only observable effect of

[22] Newton, I. (1960), p 26.
[23] Hooke, R. (2003), p 54.

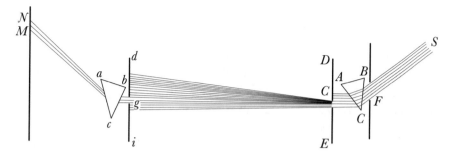

Fig. 6.5 Newton's *experimentum crucis.* Sunlight, S, enters a darkened chamber on the right and is refracted by an inverted prism ABC. A narrow portion of the resulting spectrum is selected and refracted by a second inverted prism abc. The result is a spot of colour seen on a screen at NM

refraction by the second prism was to further refract the beam of coloured light without changing its colour. Furthermore, the colour that was refracted most by the first prism was also refracted the most by the second one and vice versa. In other words, each colour was associated with a unique degree of refraction (Fig. 6.5).

Newton claimed that these observations could be explained only by assuming that white light is itself composed of "rays differently refrangible" and so is not homogeneous, as everyone at the time supposed. Moreover, he pointed out, although the number of such rays is indeterminate, we perceive them as clusters of a limited number of distinct colours, which led him to claim that the *experimentum crucis* shows that white light is made up of all the colours that are present in its spectrum. The question of why an indeterminate number of rays is not perceived as an equally indeterminate number of distinct colours was not adequately answered for another two centuries, as we shall see in Chap. 10.

And what are these colours? In his Cambridge lectures, Newton had identified five spectral colours: "red, yellow, green, blue and purple together with all the intermediate ones that can be seen in the rainbow". Now he claimed that there are seven "The Original or primary colours are, Red, Yellow, Green, Blew, and a Violet–purple, together with Orange, Indico [sic], and an indefinite variety of Intermediate gradations."[24] The addition of orange and indigo, a dark blue that few people are able to see in the spectrum because the human eye discriminates well between colours at the (brighter) red end of the spectrum but poorly at the (darker) blue end, was prompted by Newton's belief

---

[24] Newton, I. (1671/2 b), p 3082.

that there is a correspondence between light and sound due to an underlying mathematical harmony that unites all nature.[25]

The source of that idea was Pythagoras, the sixth century B.C. philosopher who famously claimed that numerical ratios are the real source of musical harmony. The faith that his followers placed in number led to extravagant yet beguiling numerological fantasies such as the Music of the Spheres, a doctrine that held that the arrangement of planetary orbits is determined by the so-called sonorous numbers that underpin musical harmony. Pythagorean numerology exerted its siren allure on many natural philosophers including Newton. Hence his assertion that sunlight is composed of seven distinct colours was not based on observation but on his belief that Pythagorean harmonies apply equally to both sound and light. It led him to claim that the spectrum of sunlight is necessarily composed of seven fundamental colours that correspond to the seven intervals of the traditional divisions of the musical octave. And to reinforce this assertion he inserted a fictitious extra colour between blue and violet and named it indigo, after the dark blue dye from India.

But his attempt to establish an objective correspondence between colour and musical harmony created a problem because, as Newton knew, colour and sound are sensations, not the inherent properties of the stimuli that give rise to those sensations. Unfortunately, he didn't make this clear to the fellows of the Royal Society at the time, which led some to suspect that he might be a closet Aristotelian who believed that colours are qualities inherent in light rather than the consequence of the interaction between matter and the sense of sight. All the mechanical philosophers of the day, Newton included, were adamant that the colours we see depend on the effect of light on our visual system.

Newton found further support for his theory that white light is composed of several colours when he discovered that pairs of simple colours in the spectrum can be combined to form intermediate colours. For example, "a mixture of Yellow and Blew makes Green; of Red and Yellow makes Orange; of Orange and Yellowish green makes yellow."[26] If simple spectral colours can be combined to form colours that to the eye are indistinguishable from those in the spectrum, why not accept that white light is also a mixture, albeit of all the colours in the spectrum rather than just two or three? In Newton's opinion this possibility was his greatest philosophical discovery.

---

[25] Should you look through a spectrometer at sunlight you will not be able to pick out Newton's indigo within the blue/violet end of the spectrum.

[26] Newton, I. (1671/2 b), p. 3082.

But the most surprising, and wonderful composition was that of Whiteness. There is no one sort of Rays which alone can exhibit this. 'Tis ever compounded, and to its composition are requisite all the aforesaid primary Colours, mixed in a due proportion. I have often with Admiration beheld, that all the Colours of the Prisme being made to converge, and thereby to be again mixed as they were in the light before it was Incident upon the Prisme, reproduced light, intirely and perfectly white, and not at all sensibly differing from a direct Light of the Sun, unless when the glasses, I used, were not sufficiently clear; for then they would a little incline it to their colour.[27]

So much for the origin of coloured light. What of the colours of solid objects? Newton found that an object's colour depends on the colour of the light with which it is illuminated.

...the Colours of all natural Bodies have no other origin than this, that they are variously qualified to reflect one sort of light in greater plenty then another. And this I have experimented in a dark Room by illuminating those bodies with uncompounded light of divers colours. For by that means any body may be made to appear of any colour. They have there no appropriate colour, but ever appear of the colour of the light cast upon them, but yet with this difference, that they are most brisk and vivid in the light of their own day-light-colour.[28]

Newton was confident that he had a watertight case. The colours in the spectrum were shown to be simple and pure, and whiteness, for so long universally associated with simplicity and purity, in optics as in human affairs, was demoted to a dependent status. He had turned Aristotle's doctrine of colour on its head. As we saw, according to this venerable doctrine, the colours of objects are real while those of lights are merely apparent. Descartes had challenged this ancient dichotomy decades before Newton. But unlike Descartes, Newton provided solid experimental evidence that light is the sole source of colour and that consequently an object's colour depends primarily on which parts of the spectrum it absorbs most strongly. Roses are red and violets are blue because red petals strongly absorb violet, blue and green light and reflect mainly red light together with some orange and yellow while violet petals strongly absorb all spectral colours except for blue and violet.

The day after the meeting, Henry Oldenburg (1615–1677), secretary to the Royal Society, wrote to Newton to inform him that his letter had "mett

---

[27] Newton, I., (1671/2 b), p 3084.
[28] Oldenburg, H. (1672). In: Turnbull, W.H., ed. (1959), pp 107–8.

both with singular attention and uncommon applause."[29] But how much of it had those present really understood its contents? They had been presented with novel ideas about light and colour, explained by appeal to experiments of which few, if any, of the assembled company had any direct experience. No experiments had been performed at the meeting because in those days the only source of light bright enough for experiments with prisms would have been the sun; Newton's paper was read during an evening meeting in the depths of winter.[30]

Even today, when Newton's ideas on the composition of white light are common currency, it helps to have a couple of prisms and a source of white light to hand as one reads through his letter. Newton did not expect anyone to take his word for his claims: "Reviewing what I have written, I see the discourse it self will lead to divers Experiments sufficient for its examination"[31] The Royal Society may have been the most prestigious scientific society in the land, indeed in Europe, but its membership was largely made up of gentlemen amateurs, not trained mathematicians or natural philosophers. So it's hardly surprising that they sought expert advice on the contents of Newton's letter before accepting its conclusions. And to whom should they turn but their "Curator of Experiments", Robert Hooke.

When the Royal Society was founded in November, 1660, the intention was to create an institution devoted to a systematic programme of research into the natural world. To this end the founding Fellows realised that they needed the services of a full-time researcher, someone who could "furnish them every day when they met, with three or four considerable Experiments." Only one man fitted the bill: Robert Hooke (1635–1703). An inventive and accomplished experimentalist, Hooke had made his name as Robert Boyle's assistant, for whom he constructed the world's first efficient air pump. This instrument, which was used to create a vacuum in a large glass vessel within which experiments could be carried out, was the particle accelerator of its time and no one was more skilled in its manufacture and use than Hooke.

---

[29] Oldenburg informed Newton that the Society had ordered his "New Theory About Light and Colours" to be printed as soon as possible in the Philosophical Transactions. It appeared in Number 80 on 19 Feb. 1671/72. It was the first research paper on a scientific topic ever to be published in a scientific journal, and marked an important turning point in the style in which scientists were to later to communicate their ideas to their peers.

[30] Newton, I. (1671/2 b), p 3084.

[31] In fact, three Fellows were asked for their opinion: Hooke, Robert Boyle, and the astronomer and mathematician, Seth Ward, who was at that time the Bishop of Salisbury. Only Hooke bothered to reply. Surprisingly, given his interest in colour, Boyle seems never to have expressed any views about Newton's theory of light and colour.

Not only was he a gifted and imaginative experimentalist, his range of interests was boundless. He was forever devising novel experiments and inventing new devices, often abandoning the project of the moment before seeing it through to completion in pursuit of something new, a trait that often tried the patience of some of the more serious Fellows.

Yet, from its inception the Society was chronically short of funds because it relied entirely on subscriptions. Although it was granted a Royal Charter by Charles II in 1662, and named the "Royal Society of London", it was an association of private individuals and received not a penny from the Exchequer.[32] Moreover, during the seventeenth century, the active membership of the society was never large—it was eighty in 1672. Attendance at the weekly meetings averaged twenty Fellows and on occasion meetings were cancelled for lack of numbers.[33] How many were present at the meeting where Newton's letter was read is not known, but it probably wasn't much more than a score. A plan to construct a building to house the Society never came to anything because the necessary funds could not be raised. As a temporary measure, between 1667 and 1673 meetings were held in Arundel House in the Strand, a private residence.[34]

Forced by its precarious finances to open its membership to gentlemen and aristocrats for the funds and prestige they might bring, Henry Oldenburg, its only full-time officer, spent much of his time chasing up unpaid subscriptions. Beset by financial difficulties the Society abandoned its plans to be a centre of research and turned its meetings into amusements for its Fellows. So if ever a man was born to be the Royal Society's Curator of Experiments, a man who could provide the Fellows with "three or four considerable Experiments at a time", that man was Robert Hooke. Goodness only knows how he found time for his other careers as a sought-after architect, land surveyor and astronomer.

Hooke was considered to be England's leading authority on the very topics that Newton had expounded to the Royal Society and had his own well-developed ideas about the nature of light and colour, which he had set forth in great detail in *Micrographia*, published by the Royal Society in 1665. In common with every other natural philosopher of the day, on the subject of light and colour Hooke was a modificationist. Like Descartes and Boyle, he believed that white light was pure and homogeneous and that the colours

---

[32] It was renamed "The Royal Society of London for the Improvement of Natural Knowledge" in 1663, when a second Royal Charter was conferred by Charles II.

[33] There were, for example, no meetings between 29th July and 30th October, 1672, when "...his Lordship [Lord Brouncker] should find a competent number of fellows in town again." See: Journal Book of the Royal Society (1672).

[34] Hunter, M. (1984).

in the spectrum are created through its interactions with matter. But where Descartes claimed that light is a pressure within ethereal matter and Boyle conjectured that it is a stream of particles moving through empty space, Hooke maintained that it is a pulse or vibration travelling through a pellucid medium that permeates both empty space and transparent matter such as air, glass and water.

According to Hooke, refraction separates the original vibration into two distinct vibrations, which he termed disturbed pulses, which travel through the refracting medium in slightly different directions. One of these pulses is responsible for redness and the other for blueness. Without bothering to explain how or why refraction gives rise to only two distinct disturbed pulses, he claimed that this was sufficient to account for the range of colours seen in the spectrum. Intermediate colours, he said, are due either to the dilution of these primary colours (for Hooke, yellow is a diluted red) or to combinations of diluted colours (green is a combination of diluted red, i.e. yellow and diluted blue).[35] Even though this account of the origin of colours is unsatisfactory, as we shall see in the next chapter, Hooke was far closer to the truth about the nature of light than either Descartes or Newton. But lacking Newton's mathematical genius and his rigorous attention to detail when conducting experiments, Hooke failed to make a coherent case for his hypothesis that the nature of light is vibrative.

Hooke was full of praise for Newton's experiments, experiments that Hooke boasted of having performed "hundreds of times", but he could accept neither Newton's hypothesis or his conclusions. And in his eagerness to find fault he failed to give Newton's experiments due consideration; indeed, he gave them short shrift. Instead, he seized upon Newton's claim that his ideas about the origin of colours and their relation to white light were founded entirely on experiment, and that no assumptions had been made about the nature of light. Nonsense, said Hooke, Newton had presupposed that light is composed of corpuscles. Furthermore, he added, Newton's *experimentum crucis* did not, in itself, support his explanation of the elongated spectrum because it did not rule out alternative explanations. Hooke claimed that his own vibration theory accounted for the results equally well, as anyone could find out by reading the relevant pages of *Micrographia*.[36] In effect, Hooke's

35 Hooke, R. (2003), pp 57–9.

36 Hooke, R. (1672). In: Birch, T. (1757), pp 10–15. During 1672, when Newton sent his letter on the nature of light and colour to the Royal Society, Hooke was busy as a surveyor working on the reconstruction of London following the Great Fire, which may go some way to explaining his curt dismissal of Newton's ideas.

assessment of Newton's claims amounted to little more than "my guess is as good as yours."

The Fellows of the Royal Society were evidently embarrassed by this cavalier dismissal of Newton's theory for they refused to publish Hooke's considerations "lest Mr Newton should look upon it as a disrespect, in printing so sudden a refutation of a discourse of it."[37] Instead a copy of the considerations was forwarded to Newton.

As if to rub salt into his wounds, as Curator, Hooke was required to perform Newton's *experimentum crucis* for the benefit of the Fellows. The Journal Book of the Society, the minutes of its meetings, records that on 18th April, 1672, "Mr Hooke was ready to make an experiment by a prism, viz. To destroy all colours by one prism, which had appeared before through another: but there being no sun, as was necessary, the experiment was deferred."[38] The first successful public performance of the *experimentum crucis* took place at a meeting of the Society a month later, on 22 May, 1672. The Journal Book entry for that day records that "Mr Hooke made some more experiments with two Prisms, confirming what Mr Newton hath written in his discovery of light and colours. Viz. That the Rays of light being separated by our Prisms into distinct colours the Refraction made by another Prism doth not alter those colours."[39] Hooke, however, continued to insist that these experiments "were not cogent to make light consist of different substances...but that this Phenomena might be replicated by the Motion of Bodies Propagated."[40]

Newton replied to Hooke's considerations by turning the tables on him, pointing out that the heterogeneity of white light could be reconciled with Hooke's own pulse theory by assuming that white light consisted not of a single vibration but of "Vibrations...of various depths or bignesses" of which the "largest beget a sensation of a Red colour, the least or shortest, a deep violet, and intermediate ones, of intermediate colours."[41] There is some uncertainty about the meaning of "bignesse" in this context. It is likely that Newton had in mind "wavelength", drawing on analogy between light and sound. He might also have been referring to amplitude.[42] But, as Newton pointed out, the greatest difficulty faced by any wave theory of light is that,

---

[37] Birch, T. (1757), p 10.

[38] Birch, T. (1757), p 43.

[39] Contrary to what you may have been taught at school, it is not possible to recombine the spectrum emerging from a prism into white light simply by passing it through a second, inverted prism unless the refracting angle of the second prism is larger than that of the first. See: Tarasov, L.V., Tarasova, A.N. (1984), p 64.

[40] Journal Book of the Royal Society, 22 May, 1672.

[41] Newton, I. (1672), p 5088.

[42] Darrigol, O. (2012), p 88.

if true, light would not be "propagated in straight lines without spreading into the shadowed Medium, on which they border."[43] And, said Newton, he knew of no circumstances in which this occurred for, as he later wrote about sounds, which were known to propagate as vibrations of air

> ...a Bell or a Cannon may be heard beyond a Hill which intercepts the sight of the sounding body, and Sounds are propagated as readily through crooked pipes as through straight ones.[44]

Christiaan Huygens, another acknowledged authority on light and optics of the time, also viewed Newton's claims with scepticism. Asked for his opinion by Oldenburg, Huygens replied that he was willing to grant that white light might be a mixture of colours, but that he felt these should be limited to as few as possible in order to make it easier to arrive at a mechanical explanation for colour.[45] Without providing any experimental evidence, he suggested that white light might be made from a combination of yellow and blue light alone. Newton countered that although it may be possible to create white light from a mixture of yellow and blue, "...such a White, (were there any such,) would have different properties from the White, which I had respect to, when I described my Theory, that is from the white of the Sun's immediate light..."[46] What he meant was that if white light is composed only of two pure colours, then refraction of such light would not produce the range of colours seen in the spectrum of sunlight. Newton could also have added that if white light composed of blue and yellow light alone is used to illuminate a surface that in sunlight appears red, then, depending on the purity of the colour of the surface pigment, the surface may appear dim and colourless, i.e. a shade of grey.

Huygens was dismayed by the bumptious tone of Newton's reply: "If therefore M Hugens would conclude any thing, he must show how white may be produced out of two uncompounded colours; wch when he hath done, I will further tell him, why he can conclude nothing from that."[47] Such condescension from a virtual unknown, which is what Newton was at the time, towards Europe's most celebrated natural philosopher was never going to lead to fruitful discourse. Huygens, who was known for his aversion to quarrels, broke off all further correspondence on the subject "...seeing that [Newton]

---

[43] Newton, I. (1672), p 5089.
[44] Newton, I. (1952), p 363.
[45] Huygens, C. (1673).
[46] Newton, I. (1673), p 6087.
[47] Newton, I., (1673), p 6087.

maintains his opinions with such warmth, I do not care to dispute the matter further."[48]

By conceding this possibility, Newton had exposed the Achilles heel of his hypothesis: his uncompromising definition of whiteness. What his prism experiments establish is that sunlight is composed of several colours, though arguably not as many as the seven he claimed to have identified. In his exchanges with Huygens, Newton made the mistake of making whiteness, rather than the heterogeneity of light, the central issue. But he took the lesson to heart: never again was he to claim that whiteness as opposed to sunlight is necessarily composed of seven colours.[49] We now know that white light can be created artificially by mixing no more than three well-chosen coloured lights. In fact, this is now commonplace: examine a white patch in an image in a television screen or computer monitor with a magnifying glass and you will see that it is made up of a multitude of tiny red, green and blue lights, not of all the colours of the classic Newtonian spectrum. Indeed, white light can be made by combining two colours, just as Huygens suggested, as long as these are strictly complementary.

The problem of interpreting the results of Newton's experiments wasn't the only thing that held back the acceptance of his ideas. Far from convincing the doubters, the *experimentum crucis* proved more of hindrance than a help because his description of this key experiment was insufficiently detailed, making it difficult for his critics—except for Hooke—to obtain the same results when they repeated it. Then there was the poor quality of the glass from which prisms and lenses were made in those days, which usually contained bubbles of air and was tinged with colour. But the greatest stumbling block was that none of Newton's contemporaries matched his experimental skill. In any case, he had described only a tiny handful of his optical experiments in his letter to the Royal Society, so no one was aware of the huge number of experiments, many of them far more elaborate and revealing than the *experimentum crucis*, that he had in fact performed, and which collectively provided overwhelming evidence that white light is not homogeneous. Although he had described and explained these experiments in detail in his Cambridge lectures, the lectures were not published in Newton's lifetime.[50]

[48] Huygens, C. (1897), p 302.

[49] Shapiro, A. E. (1980), p 224.

[50] They were published in 1728 under the title of "Optical Lectures read in the publick schools of the University of Cambridge, anno Domini, 1669". For a modern edition of the lectures: Shapiro, A.E. (ed) (1984).

Newton was initially prepared to answer his critics, sometimes at length, but grew increasing exasperated when he realised that he wasn't going to allay their doubts and convince them of his case.[51] As far as he was concerned, he had prised open Nature's paintbox with a wedge of glass to reveal her hidden pallet of elemental colours. If the best minds of the age wouldn't accept the evidence that he had done so, so be it; it was time to break off all further correspondence on the subject. In June 1673 he wrote a final petulant letter to Henry Oldenburg, who had acted as Newton's go-between with his many correspondents:

> I intend to be no further solicitous about matters of Philosophy. And therefore I hope you will not take it ill if you find me ever refusing doing any thing more in that kind.[52]

But Newton didn't cut himself off completely. In 1675 he sent a long letter to the Royal Society on the subject of colours seen in thin transparent films in which he reluctantly committed himself to explaining the physical basis of the phenomenon "assuming the rays of light to be small bodies emitted every way from shining substances".[53] The only reason he was prepared to say even this much about the nature of light, he added irritably, was because "I have observed the heads of some great virtuosos to run much upon hypotheses, as if my discourses wanted an hypothesis to explain them by, and found, that some, when I could not make them take my meaning, when I spake of the nature of light and colours abstractedly".[54] But this paper made no mention of the rainbow because Newton thought that he was dealing with a phenomenon that played no part in its formation. We'll see how wrong he was in the course of the next two chapters.

At Newton's request, this letter was not published in the Philosophical Transactions and so, bearing in mind how few Fellows attended meetings at the Royal Society and that most of those were dilettantes, the world at large had to wait until 1704, when he published *Opticks, or a Treatise of the Reflections, Refractions, inflections and Colours of Light*, for a full account of his ideas on light and colour and to learn of the scores of revealing experiments that he had not mentioned in his original letter to the Royal Society.[55]

---

[51] Newton had supporters for his theory of colour in the Royal Society, and he also managed to convince one of his European critics, the Jesuit natural philosopher, Ignace Pardies (1636–1673). See: Sabra, A.I. (1981), pp 264–8.

[52] Newton, I. (1673). In: Turnbull, W.H. (1959), pp 294–5.

[53] Newton, I. (1675), p 249.

[54] Newton, I. (1675), p 249.

[55] Newton, I. (1952).

Following his 1675 letter, Newton turned in on himself and avoided all dealings with the wider world where possible. As Lucasian Professor he continued to deliver mathematical lectures, but otherwise showed little appetite for or interest in either mathematics or natural philosophy. For the next ten years alchemy and theology occupied him to the exclusion of almost everything else.

He was, as might be expected of someone of his time and place, a pious and God-fearing Puritan. But such a driven man could never have been expected to take the teachings of the Church at face value, for he made no distinction between theology and natural philosophy, holding that both offered insights into God's plan. His exhaustive study of Christian literature convinced him that the doctrine of the Trinity—the idea that God, Jesus and the Holy Ghost are one and the same, which had been established during the late fourth century AD and became one of the central doctrines of both the Catholic Church and the of Church of England—was a heresy perpetrated by some of the founding fathers of the Catholic Church. Newton disputed the divinity of Christ, and sought evidence for his views in the scriptures, which he maintained had been subtly rewritten to support the Trinitarian doctrine. He taught himself Hebrew in order to study the original texts of the Bible and pored minutely and obsessively over vast numbers of arcane manuscripts almost to the day he died and did, indeed, discover that the Trinitarian doctrine was not present in the earliest Hebrew versions of the Bible. But this opinion was a heresy in seventeenth century England, and of the millions of words he wrote on theological matters only 850 pages of his The Chronology of Ancient Kingdoms Ammended, described by one of his biographers as "a work of colossal tedium", was published, and posthumously at that.[56]

He was careful to keep his religious beliefs to himself, knowing that they would harm his career if they became known. Fortunately for Newton—and for science—the Lucasian Professorship was the only academic post in Cambridge that did not require its incumbent to take holy orders on appointment. Had it done so, Newton would not have been appointed, for it is most unlikely that he would have been prepared to set aside his religious scruples concerning the Church of England. And without the tenure of a university post how would an impecunious country lad, however gifted, had an opportunity to pursue his researches?[57]

The penchant for bookish pedantry that served him so well in his biblical studies had other outlets. Allied with his love of experiment it enabled him

---

[56] Westfall, R.S. (1980), p 815.

[57] Many of the leading natural philosophers of the 17th C were either independently wealthy (Descartes, Boyle and Huygens) or had patrons (Galileo).

to pursue an even greater passion. Newton, the leading mathematician and natural philosopher of his day, was also one of its foremost alchemists. During his long tenure as Lucasian Professor, Newton spent as much, if not more time familiarising himself with all the available alchemical literature and with the unrewarding business of the transmutation of elements as he did on mathematics and natural philosophy. He found alchemy a confused and disparate discipline and set about making sense of it. It seems more than likely that his readiness to countenance the decidedly non-mechanical ideas of forces acting between bodies that are not touching one another—most famously that of gravitational attraction—owes something to his immersion in alchemy with its notions of occult powers invested in matter.

The purpose of Newton's original letter to the Royal Society had been to provide the background to the design of his reflecting telescope. Despite the interest it aroused, the telescope didn't live up to its promise. The main problem was that the mirror was made of a metal alloy, not glass. The alloy proved too soft to shape accurately and it tarnished easily. Even when freshly polished, the mirror reflected poorly, producing a murky image. Newton didn't use glass because at the time it wasn't known how to silver the surface of a glass mirror, necessary in a reflecting telescope because it eliminates the multiple images that are produced in mirrors that are silvered on the back.[58]

The advantage of using a mirror in a telescope is that, unlike refraction through a lens, reflection from a smooth surface does not lead to a separation of colours. So, because the light-gathering element in Newton's telescope was a concave mirror and not a lens, its design didn't depend on knowing the degree to which each colour refracts. It was enough to know that red light is not refracted quite as much as violet light to see the advantage of such a telescope over one that employs lenses.

But in the absence of experiments with prisms, the only really convincing example that white light is a mixture of colours that Newton had to offer the audience at the Royal Society on the Wednesday evening when his letter was read out was his explanation for the order of colours in the primary and secondary rainbows. Unlike the design of the reflecting telescope, this does require a precise knowledge of the degree of refraction of each colour. More importantly, since few among the Fellows would have seen a spectrum produced by a prism under the rigorous conditions of Newton's experiments, let alone have had a mental picture of one, the rainbow was probably the only example of the consequences of his ideas about white light with which they would all have been familiar (Fig. 6.6).

---

[58] The first surface-silvered glass mirror was made in 1856. The technique used was based on an 1835 discovery by the German chemist, Justus von Liebig (1803–1873).

Fig. 6.6 Colours that are seen in a bright rainbow are less bright and well defined when compared to those seen the spectrum formed when sunlight is dispersed by a prism[59]

Nevertheless, on that occasion he confined his remarks on the rainbow to a single paragraph.

Why the Colours of the Rainbow appear in falling drops of Rain, is also from hence evident. For, those drops, which refract the Rays, disposed to appear purple, in greatest quantity to the Spectators eye, refract the Rays of other sorts so much less, as to make them pass beside it; and such are the drops on the inside of the Primary Bow, and on the outside of the Secondary or Exteriour one. So those drops, which refract in greatest plenty the Rays, apt to appear red, toward the Spectators eye, refract those of other sorts so much more, as to make them pass beside it; and such are the drops on the exterior part of the Primary, and interiour part of the Secondary Bow.[60]

---

[59] Photo by the author.
[60] Newton, I. (1671/2 b), p 3084.

This was but a brief of summary of the detailed mathematical investigation into the rainbow that he had included in his Cambridge lectures and which was later published in his *Opticks*.[61]

In one of those lectures, Newton had taken up the explanation of the rainbow where Descartes' had left off. We have seen how, using the laws of refraction and reflection, Descartes had worked out the exact mathematical relationship between the angles of refraction and reflection for a ray of light through a drop and that he laboriously calculated the path of several individual rays, eventually stumbling across those responsible for the rainbow. Newton took the relationship that Descartes had established between the various angles and applied to it the mathematical technique that he had recently developed, the calculus. This enabled him to determine the smallest angle through which a ray of light is deviated by the raindrop, in other words, the angle of least or minimum deviation, thereby identifying the ray that marks the outer edge of the rainbow, the so-called rainbow ray, without the need for numerous calculations.

In the last of his lectures, Newton had tackled the rainbow's colours. Knowing that red light refracts slightly less than violet light, Newton understood that the path of the least refrangible rays through a drop—i.e. those at the extreme red end of the spectrum—must differ slightly from those of the most refrangible rays, those at the extreme violet end. But to carry out calculations to determine the paths he had first to measure the refractive index of these rays in water and, being methodical to a fault, did so for rainwater. Unfortunately, with the apparatus at his disposal, the extreme ends of the spectrum were faint and fuzzy, forcing him to make an arbitrary decision about the rays that undergo the greatest and least amount of refraction. Nevertheless, the values for refractive index he obtained for red and violet light are not far off those accepted today.

Using these values and his calculus-derived formula for the least-deviated rays, Newton was able to calculate the smallest angles at which light from the outermost edge of red and violet ends of the spectrum emerge from a drop after one and two internal reflections, thus showing that the observed widths of the primary and secondary bows and the order of colours within each of them are a direct consequence of the fact that white light is a mixture of colours, each of which is associated with a particular degree of refraction.

Because the Cambridge lectures were not published during Newton's lifetime, these calculations became known only when he included them in the explanation of the rainbow given in *Opticks*.[62] This consists mainly of rather

---

[61] Newton, I. (1984).

[62] Newton, I. (1952), pp 168–78.

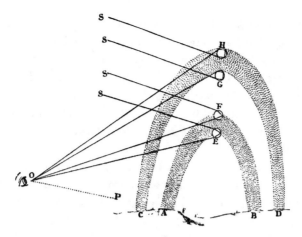

Fig. 6.7 Newton's rainbow. Ray SFO is responsible for the red band of the primary bow, SEO for the blue/violet band. Ray SGO gives the red band of the secondary bow, SHO the blue/violet band. The observer is at O, the antisolar point at P[63]

forbidding computations of the angular sizes of the primary and secondary bows and adds nothing to what he had to say on the subject in his lectures. According to these calculations, the outer radius of the primary bow, due to rays that "strike the Senses with the deepest red Colour", is 42° 2 min. The inner radius of this bow is 40° 17 min and is due to rays "with the deepest violet Colour". The corresponding values for the secondary bow—which is due to rays that are reflected twice within a drop—he made 54° 7 min and 50° 57 min respectively. And due to that second reflection, the colours are reversed so that the inner edge of the secondary bow is red and the outer edge is violet (Fig. 6.7).

As they stand, these calculations assume that the sun's light comes from a single point on its surface, which is not correct. Allowance must be made for the sun's apparent diameter, which is half a degree, or thirty minutes of arc. Taking this into account Newton concluded "the breadth of the interior Iris will be 2° 15 min that of the exterior 3° 40 min" and that the gap between them, Alexander's dark band, is "…8° 25 min…" . We shall return to these calculations in the next chapter.

This was not quite Newton's last word on the subject of the rainbow. He pointed out that there should be a third rainbow, though perhaps it might not be bright enough to be visible because "The light which passes through a drop of Rain after two Refractions, and three or more Reflections, is scarce strong

[63] Newton, I. (1952), Fig. 15, p 173.

enough to cause a sensible Bow."[64] Newton was content merely to mention the possibility of what has come to be known as the tertiary bow in *Opticks*, though he had long before calculated where it might be seen and included this information in his Cambridge lectures on optics.[65] The lectures were not published until 1728, and it was the affable and very able Dr Edmond Halley (1656–1742) who was first to publish a detailed study of rainbows due to multiple reflections and thus round off Newton's account of the rainbow.[66]

Halley is known to most people as the astronomer who identified the brightest of all periodic comets, later named in his honour.[67] Though he is remembered chiefly as an astronomer, he was a man of wide interests and a first-rate mathematician, admired as such throughout Europe. He made his reputation in his early twenties when he published a star atlas of the Southern skies based on systematic observations he had made from the island of St Helena in 1677, having abandoned his studies at Oxford University to do so. During his long life he was active member of the Royal Society—he was made a Fellow in recognition of his star atlas soon after his return from St Helena—and a pioneer in several new sciences including geophysics and actuarial science. Although it is possible to form an opinion about his personality through his work and by reading between the lines of his many papers and letters, from which he emerges as a remarkably likable and well balanced individual, about the only report we have that gives a sense of what he was like in the flesh is that "in his person he was of a middle stature, inclining to tallness, of a thin habit of body, and a fair complexion, and always spoke as well as acted with an uncommon degree of sprightliness and vivacity."[68]

He also was a born diplomat, a quality that proved invaluable when dealing with some of the prickly and peevish personalities he had to work with in his professional life. Thus, for all his success as an astronomer, mathematician and natural philosopher, not the least of his achievements was to bring Newton back to natural philosophy when, in 1684, he travelled to Cambridge with the express purpose of obtaining an answer to the question that was to wake Newton from his scientific slumber: how to explain Kepler's discovery that the orbit of a planet about the sun is elliptical, not circular.

---

[64] Newton, I. (1952), p 178.

[65] Newton, I. (1984), pp 423–5.

[66] Independently of Halley, Johann Bernoulli (1667–1748), the Swiss mathematician, also calculated the location of the tertiary bow. See: Bernouilli, J. (1742), Vol IV, pp 197–203.

[67] A periodic comet is one that orbits the sun over and over again. There are several such comets, of which Comet Halley is by far the brightest and best known. It returns to our skies once every 75 years. In 1705 Halley predicted the comet would reappear in 1758. He did not live to see his prediction confirmed. It is due to return to our skies in 2061.

[68] Biographia Britannica (1757), p 2517.

In fact, Newton already had the beginnings of an answer, which he had probably formulated when he first began to think about gravity, supposedly after seeing an apple fall to the ground in the orchard at Woolsthorpe Manor some twenty years earlier. Halley's visit reawakened Newton's interest in the issue and galvanised him into action. The fire in his alchemical furnace was allowed to go out and the arcane theological texts were returned to their shelves as he struggled to reformulate his ideas on matter and motion and revisited unfinished mathematical projects abandoned years previously. For the next two years, the task consumed him to the exclusion of everything else and resulted in his masterwork, by common consent the most important and influential scientific book ever written: *Philosophiae Naturalis Principia Mathematica*.[69]

There are almost no first-hand accounts of Newton during his time at Cambridge. Fortunately, one of the very few we have coincides with the period he was working on the *Principia*. We owe it to Humphrey Newton, his assistant for some five years during the late 1680s. Humphrey Newton, who was not a relative, assisted Newton in his alchemical experiments and made a fair copy of the *Principia* in preparation for its publication. He has left us with a portrait of the quintessential absentminded professor, so thoroughly engrossed in his work that "When he has sometimes taken a Turn or two [about his garden], has made a sudden stand, turn'd himself about, run up ye stairs [and] fall to write on his Desk standing, without giving himself the Leasure to draw a Chair to sit down in." As for his person, "…shoes down at Heels, socks untied … his Head scarcely combed", "…so serious upon his studies that he eat very sparingly…" and very rarely went to bed "till 2 or 3 of the clock…".[70] Such are the outward signs that ordinary folk associate with a genius absorbed in the act of creation.

Having coaxed Newton into writing the book, Halley then acted as its patient and forbearing editor. Although the work was originally commissioned by the Royal Society, when it came to publication Halley was obliged to bear the cost of printing it because the Society had squandered its funds on the publication of a lavishly illustrated edition of *The History of Fishes* by the naturalist, Francis Willoughby (1635–72). Willougby's book failed to sell, and was instead used to reimburse Halley for his pains as the Society's Clerk. Happily, it seems that Halley was not left out of pocket by underwriting the cost of publishing the *Principia* and may even have made a modest profit

---

[69] Newton, I. (1960). Newton chose the title as a snub to Descartes' *Principia philosophiae*. Hence "*The Mathematical Principles of Natural Philosophy*" as distinct from Descartes' "*Principles of Philosophy*".
[70] Newton, H. (1727/28).

from the sale of the 500 or so copies of its first edition. Newton received twenty copies "to bestow on [his] friends" and a further forty to "to put into the hands' of Cambridge booksellers."[71]

The serendipitous sequence of events that led Newton to write *Principia* illustrates how slender is the thread by which hangs a man's reputation and influence. Without Halley's involvement Newton would almost certainly have remained lost in a world of his own, pursuing the chimera of ancient wisdom that he believed was locked away in the esoteric literature of alchemy and in ancient religious texts. Halley's timely intervention rescued Newton from posthumous obscurity and transformed the little-known university professor into the Promethean figure who almost single-handedly laid the foundations of modern mathematical physics. Had Newton died in 1684 (aged 42 years), posterity might have acknowledged his genius but concluded that he had not lived up to his early promise. Even though Newton lived to a ripe old age, without Halley not only would *Principia* not have been written, neither would *Opticks*. As for the subsequent development of science, especially physics, in the absence of Halley's fortuitous intervention we can only speculate, for in his day Newton had no equals either as a mathematical physicist or an experimentalist. If ever there was an example of cometh the hour, cometh the man, where science is concerned, Newton is surely it. Or should that be Halley?

Halley shared Newton's fascination with the rainbow. His interest in rainbows appears to have been whetted the year before he saw the rainbow due to reflected sunlight when he was in Chester, described in Chap. 2. In the letter to the Royal Society describing this rainbow he mentioned that a couple of years earlier, in 1696, he had been walking up Abchurch Lane in the City of London during a rainstorm when he saw a rainbow spanning the street like an

...arch of a building, under which I was to pass; the crown whereof was not much higher than my Head, and the diameter thereof scarce so wide as the Street, which is but 5 Yards; and it moved along with me as I went; the Colours being very vivid and distinct, though the Arch itself appeared but narrow, and Houses were everywhere behind it. This, tho' very uncommon, will not appear strange to those that have well considered the Nature of the Iris; but the Ancients who believed Iris the messenger of the gods, would have been apt to have thought she had some peculiar message, when she placed herself so near me, as to be almost within reach: I understood her to invite me to inquire further into the Nature of her Production and accordingly, taking

---

[71] Feingold, M., Svorenčík, A. (2020), pp 253–348.

her under my Consideration, I had all the Success I could wish for, Which perhaps may not be unacceptable to the Curious, if I publish in one of the next Transactions.[72]

A couple of years later, in 1700, he made good his word and published a general account, written in Latin, of rainbows due to repeated reflections within a drop.[73] Following Newton's lead, he employed calculus to determine where each bow would be seen. He calculated that a tertiary rainbow would be about the same size as the primary bow, i.e. have the same angular radius, though its arc would be twice as broad, and that it would be centred on the sun so that one would have to face the sun to see it. And, as with the primary bow, the inner arc is violet and the outer one red. However, because it is due to three reflections within a drop, the tertiary bow is extremely faint and so is virtually impossible to see against the glare of the sunward sky. Reports of its sighting have been few and far between, so until very recently it was difficult to judge if this bow is actually visible in natural surroundings. What has tipped the balance in favour of sightings is that tertiary rainbows have been photographed on several occasions so there can be no doubt that given favourable conditions it really is visible.[74]

Halley went on to determine the theoretical size and position of the rainbow due to four internal reflections, known as the quaternary rainbow, and found that its arc is slightly wider than the tertiary bow.. The quaternary bow also forms around the sun, and, like the tertiary bow, its violet edge is further away from the sun than its red edge. It is, of course, even fainter than the tertiary bow, though it too has been photographed.[75] It can be seen under laboratory conditions because this allows one to reduce background brightness to the point where it becomes visible, as do bows involving five, six and more reflections. We'll have more to say about these "higher order" bows in Chap. 10 (Fig. 6.8).

Halley also dealt with unfinished business in Descartes' account of the rainbow. Descartes had ended his explanation of the rainbow on a puzzling note with a brief description of an invention "…for making signs appear in the sky, which would cause great wonder in those who were ignorant of the causes."[76] This was a fountain fed with "…oils, spirits and other liquids, in which refraction is notably greater or lesser than in common water…"

---

[72] Halley, E. (1698), p 195.
[73] Halley, E. (1700), pp 714–25.
[74] Haußmann, A. (2016), p 37.
[75] Theusner, M. (2011).
[76] Descartes, R. (2001), p 344.

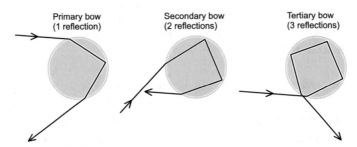

Fig. 6.8 The path of the rainbow ray within a raindrop responsible for the primary, secondary and tertiary rainbows. Note that the rainbow ray responsible for the tertiary bow emerges from a drop in the same direction as that of the light entering the drop

creating a series of sprays in which one would see "…a great part of the sky full of the colours of the rainbow." The purpose such a fountain is difficult to fathom; on the face of it, the wish to surprise rather than enlighten seems a betrayal of everything Descartes stood for. But perhaps the idea of such an outlandish device sprang from his belief that wonder is the spur that leads one to seek knowledge and understanding. The wonder engendered by the spectacle of this peacock's tail of individual bows of different diameters might encourage the spectator to inquire into its cause.

Descartes may have been aware that the radius of a rainbow depends on the refractive index of the substance of the drops in which it is seen, but it was Halley who carried out the necessary calculations. He showed that as the refractive index of the substance of which the drops are composed increases, the resulting bows change size, with the primary bow becoming smaller and the secondary bow growing larger. These calculations revealed that if raindrops were composed of diamond rather than of water, there would be no primary bow and only the secondary bow would be visible. As far as actual rainbows are concerned, all this is largely academic because, apart from water, there are few substances in nature that occur in form of a large number of small, transparent spheres. The only exception is seawater, which is slightly more refractive than fresh water. As a result, the angular radius of a bow formed in seawater spray is approximately 1° less than that formed in fresh water, but this is something that you are only likely to notice when both bows are seen at the same time.[77] Bows can also be seen in surfaces covered in tiny glass spheres. Tiny glass spheres are sometimes used in road signs or added to paint used in road markings to make them brighter. Taking the refractive

---

[77] Photo of a seawater bow together with a rainwater bow: https://tinyurl.com/2p9yt29y. Accessed 30/06/22.

Fig. 6.9 A puddle of colour. A fragment of a rainbow formed by glass beads on a street sign and reflected in a puddle[79]

index of glass to be 15% more than that of water, the resulting primary bow has a radius of approximately 25°, approximately half that of a bow formed in drops of water (Fig. 6.9).[78]

Halley didn't carry out any experiments to demonstrate the effect on the dimensions of a bow of the substance in which it is formed, but in 1884, John Tyndall (1820–1893), an Irish physicist famed for his optical researches, created spraybows using different liquids at the Royal Institution in London where he was its professor of physics and Michael Faraday's successor as its Director. He began with water followed by turpentine, which produced a "circular bow of extraordinary intensity", paraffin, chloroform, benzine and carbon tetrachloride, to name a few of the more hazardous of the many substances he experimented with.

Tyndall also created Descartes' rainbow fountain: "Having produced the extremely vivid bow of the turpentine shower, the water was turned on. Its

---

[78] Christiaan Huygens had calculated the radius of a rainbow formed in glass beads in 1652. He made it 21°45′, which given that he took the refractive index of water to be "250/187", is far too small. Despite being urged to publish this work, it didn't see the light of day until 1703, several years after his death. See: Huygens, C. (1916), p xiv and pp 10–13.

[79] Photo by the author.

spray fell, mingled with the turpentine, and, instantly, round the turpentine bow the larger water-bow was swept through the darkness."[80] For anyone tempted to carry out similar experiments at home, Tyndall recommended "the sprays of turpentine and petroleum", adding helpfully "I need not say that caution is necessary in dealing with inflammable liquids."[81] One of his light sources was a naked lime light, which was separated from the spray by a plate of glass. The experiments confirmed Halley's calculations.[82]

Newton's account of the rainbow in *Optiks* differed from that in his lectures in one notable respect. In his Cambridge lectures Newton had quite rightly given Descartes all the credit for explaining of the size of the rainbow. But almost from the day he had first encountered Descartes, Newton was plagued by the thought that the Cartesian separation between mind and matter opened the door to atheism, something he viewed with alarm. Newton never could imagine a universe that was not the creation of a deity. Nor could Descartes; but where Descartes believed that God had made the universe, devised the laws of nature that would govern all events that take place in it, and then left it to its own devices, so to speak, Newton believed that God was ever-present and from time to time intervened to correct the workings of nature such as adjusting planetary orbits when necessary.

By the time he came to write *Opticks*, Newton had developed such an antipathy towards Descartes that he could hardly bring himself to write his name.[83] And in later editions of *Opticks* he took the opportunity to belittle one of Descartes' few scientific successes that has stood the test of time by claiming that Descartes had slavishly copied the ideas of "the Famous Antonius de Dominis Archbishop of Spilato".[84]

According to Newton, the only aspect of the rainbow that Dominis had not succeeded in explaining was its colours. Yet he must have known that Dominis had failed to provide an adequate explanation for any of the rainbow's features, for he possessed a copy of Dominis' book, *De Radiis Visus Et Lucis*. A glance at the crude figures in this book shows all too clearly that Dominis' account of the passage of light through a spherical drop is far inferior to that of Theodoric, let alone to that of Descartes, because Dominis ignored the refraction that occurs as light emerges from the drop. At the same time, the figure muddles up the rays responsible for the primary and secondary bow (Fig. 6.10).

---

[80] Tyndall, J. (1884), p 64.

[81] Tyndall, J. (1884), p 64.

[82] As far as I am aware, no one has repeated Tyndall's experiments.

[83] Westfall, R.S. (1980), p 401.

[84] Newton, I. (1952), p 176.

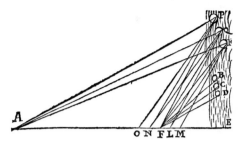

Fig. 6.10 de Dominis' rainbow. The primary bow is due to light from drops BCD, the secondary bow is due to light from drops PQF. A is the sun and F the position of the observer.

The diagrams also show that Dominis had not freed himself from his Aristotelian roots: light from the sun diverges from a point on the horizon and is redirected towards the observer by raindrops high in the sky without being refracted within drops. Furthermore, he explained the rainbow's colours as the weakening of light due to reflection.

Unfortunately for Descartes' claim to be the author of the first correct account of the size and shape of the rainbow, few of Newton's readers had access to *De Radiis Visus et Lucis* and so gave Dominis the benefit of the doubt on Newton's say-so. Astonishingly, in the English-speaking world, if nowhere else, Newton's libel against Descartes was still being repeated in textbooks well into the twentieth century.[85] It is time to put the record straight: Descartes' account of the size and shape of the rainbow is entirely original. He probably knew nothing of Dominis' explanation of the rainbow. Even if he had read Dominis' little book on the subject, Descartes, like everyone else, would have leaned almost nothing of value from it.[86]

[85] For examples of this scientific libel see: Preston, T. (1912), p 548, Armitage, A. (1950), p.4, Hulst, H.C. van de (1957), p 240.

[86] J.-P. Biot made this point in the early 19th C. See: Biot, J.-P. (1826), p 175.

7

New Wine in Old Bottles

But rainbows differ, among themselves, as one tree from another, and, besides, some of their most interesting features usually are not even mentioned—and naturally so, for the "explanations" generally given of the rainbow may well be said to explain beautifully that which does not occur, and to leave unexplained that which does.[1]

Between them, Descartes, Newton and Halley had taken the explanation of the rainbow as far as possible given what was known about light and colour in their day. Not that they or anyone else at the time had the slightest inkling that there was more to explain. Doubts would only begin to surface long after they were all dead. At the time it seemed as if their explanations were both comprehensive and unassailable. Descartes had shown that a rainbow is a narrow, circular arc whose size depends on the way that light is refracted within a transparent sphere to form a bright caustic surface. Newton had explained how its colours and the order in which they are seen are a consequence of the fact that light is a mixture of rays "differently refrangible". And Halley had completed Newton's account by working out the optics of a reflected rainbow, where to look for the yet to be seen third rainbow and the effect on the appearance of a rainbow formed in drops of substances other than water.

---

[1] Humphreys, W. (1929), p 458.

© The Author(s), under exclusive license to Springer Nature
Switzerland AG 2023
J. Naylor, *The Riddle of the Rainbow*, Copernicus Books,
https://doi.org/10.1007/978-3-031-23908-3_7

But for all its success, there is no getting away from the fact that Newton's explanation of the rainbow, like Descartes', is largely a pencil and paper exercise that relies more on geometry than physics. He had little choice, for despite bold claims to the contrary, every theory about the nature of light at the time was based on conjectures for which there was little or no experimental evidence. As we have seen, during the seventeenth century several new theories about the nature of light were proposed. Broadly speaking these fell in two camps. The Cartesians maintained that light is a disturbance within a medium—known as the æther—that fills all space, an idea that was taken up and refined, first by Robert Hooke and later by Christiaan Huygens. The Newtonians—of whom, it must be said, there were very few, if any, until the publication of *Principia*—held out against these Cartesian notions and resolutely maintained that light consists of a stream of particles. Ironically, as we shall see later in this chapter, one of the few scientists of the time wary of committing himself wholeheartedly to any theory about the nature of light, including his own, was Newton himself.

Despite Newton's equivocation, the first round went to the Newtonians, and for more than a century following the publication of his *Opticks* in 1704 few natural philosophers questioned what they took to be Newton's final thoughts on the matter. They overlooked or ignored the fact that time and again he had refused to be drawn on the nature of light. When Hooke claimed that a corpuscular theory of light was essential to Newton's theory of colours, Newton was quick to distance himself: "'Tis true, that from my Theory I argue the Corporeity of Light; but I do it without any absolute positiveness, as the word perhaps intimates; and make it at most but a very plausible consequence of the Doctrine, and not a fundamental Supposition".[2] But Newton was not being altogether frank when he wrote this, for he always maintained that the fatal weakness of all theories of light based on waves or pulses was that they require that light should wrap around objects as waves in water do. The fact that a shadow of an object illuminated by a point source of light has a sharp edge implies that light travels in straight lines.[3] The only viable explanation for this fact seemed to him to be one based on corpuscles. As he put it years later in *Opticks*, if only rhetorically.

---

[2] Newton, I. (1672), p 5086.

[3] A shadow in sunlight has a fuzzy edge (known as the penumbra) because the sun is not a point source so illuminates an edge from several directions at once.

Are not the Rays of Light very small bodies emitted from shining Substances? For such Bodies will pass through uniform Mediums in right Lines without bending into the Shadow, which is the Nature of Rays of Light.[4]

Nevertheless, Newton's preference for a corpuscular theory of light amounted to little more than window dressing got up to appear as if it was the only feasible interpretation of carefully considered experiments. But because Newton's optical experiments were so exquisitely designed and executed, his corpuscular theory became an article of faith for most natural philosophers in the eighteenth century, while promising alternatives, such as Huygens' mathematical version of the wave theory, were ignored.

In the absence of any concrete knowledge about the nature of light, Newton and Descartes were fortunate that in many circumstances the effects of reflection and refraction can be adequately described while ignoring the nature of light altogether and simply representing these effects by lines that plot the path taken by a narrow beam light, an approach known as geometrical optics. Geometrical optics provides a perfectly acceptable means with which to explain a huge number of optical phenomena, including many of the rainbow's salient features, but since it ignores the actual nature of light, it can't account for effects that are due to light's wave nature.

Although Newton repeated Descartes' experiments with a glass flask, he did so to confirm what he already knew about rainbows rather than as an investigation to discover new facts about them. We would expect nothing less of such an inveterate experimentalist than the need to check things for himself. As we saw in the previous chapter, the only novel element in his account of the rainbow was to employ his discovery that colour is associated with the degree of refraction, something that Descartes had overlooked. And Newton, unlike Descartes, did at least compare the results of his calculations with real rainbows and so was aware of the difficulty of doing so, even though he brushed the significance of the difficulty aside.

For however elegant and persuasive any explanation of the rainbow seems on the page, the only acceptable measure of its success is how well it agrees with observation. On the face of it, Newton appears to have accounted for every feature of the rainbow. Not only did he confirm Descartes' estimate for the theoretical maximum radius of the primary bow (the outer edge of the red segment of the primary arc) and the minimum radius of secondary (the inner edge of the red segment of the secondary arc), but by taking the heterogeneous nature of light into account he was also able to account for their several colours, the order in which they are seen, and in the process

---

[4] Newton, I. (1952), p 370.

establish a theoretical value for the width of the primary and secondary arcs. But little else: his explanation certainly doesn't prepare one for the fact that no two bows are identical.

In the first place, the precision with which Newton—and for that matter, Descartes—calculated the dimensions of the rainbow should be taken with a pinch of salt because, as Newton discovered when he compared his theoretical rainbow with the real thing, both the inner and outer edges of natural bows are never sharply defined. On one occasion he found that the inner violet edge of the primary bow "was so much obscured by the brightness of the Clouds, that I could not measure its breadth."[5] Nor did the fact that the outermost band was a "faint red" help matters for it made it impossible to determine the outer edge of the bow precisely.

Nevertheless, despite these difficulties, Newton was prepared to stick his neck out and state unequivocally "such are the Dimensions of the Bows in the Heavens found to be very nearly, when their colours appear strong and perfect."[6] He was equally unequivocal about the colours in the primary arc: "from the inside of the bow to the outside in this order, violet, indigo, blue, green, yellow, orange, red", though added that "the violet, by the mixture of the white Light of the Clouds, will appear faint and incline to purple."[7] The same colours, he said, are present in the secondary bow though they "lie in the contrary order" and are much fainter than in the primary bow because they are due to light that has been reflected twice within a drop, for, as he correctly pointed out, "Light becomes fainter by every Reflexion."[8] The reason this occurs is that that water is transparent and so at the point within a drop at which it is reflected most of the light exits to the surrounding air. Only a tiny fraction of light is reflected, which means that the second reflection consists of a fraction of a fraction of the original beam that originally entered the drop.

Unfortunately, the account of the rainbow given in *Opticks* is still widely taken to be accepted the last word on the subject, which it emphatically is not. When photographs of rainbows are compared, it is obvious that not only does the brightness of their colours vary from one rainbow to the next, and even within a single bow, but also that colours in a rainbow are never as vivid or as pure as the colours produced by a prism. Moreover, some of the colours listed by Newton are not always present. Newton's present-day acolytes have some explaining to do.

---

[5] Newton, I. (1952), pp 175–6.
[6] Newton, I. (1952), p 175.
[7] Newton, I. (1952), pp 173–4.
[8] Newton, I. (1952), p 177.

In fact, Newton could have accounted for some of these observed discrepancies had he wished to do so, but perhaps he felt that having successfully explained the main features of the rainbow someone else would deal with these seemingly minor details. Whatever the reason, as his remark about the effect of cloud-light on the appearance of the violet band in a bow demonstrates, he had an inkling that when looking at a rainbow what we see is always a mixture of light from the rainbow itself and light from the background against which it is seen. Indeed, the only circumstances in which we would see a rainbow's actual colours, unmixed with those of its background, is if it were seen against a completely black background; the best that nature can offer is a dark storm cloud.

Even under the most favourable circumstances, however, a rainbow's colours are never pure because neighbouring colours slightly overlap one another. The distinct and vivid colours seen in a spectrum created under the laboratory conditions of Newton's *experimentum crucis* can only be produced by passing a narrow, parallel beam of light through a prism.[9] Sunlight, of course, is not parallel because the sun is a disc, not a point like a star—this is why shadows cast in sunlight have a fuzzy edge, known as a penumbra. Consequently, each colour spreads out as it emerges from a raindrop in the direction of the eye and mixes with others to some degree. Thus light at the edge of the yellow band is mixed with some light from the red one, that of the edges of the green band is mixed with both some blue and yellow light and so on. Within the area enclosed by the primary bow the colours all mix to form white light. A rainbow is decidedly not the paradigm of nature's paintbox that we are led to believe. At best it is a pale, confused and inconstant display of sunlight's elemental colours.

The impurity of the rainbow's colours isn't the only issue that Newton's account fails to address. Both he and Descartes knew that in a rain shower there are drops of many different sizes.[10] What they would not have known is that there are always many more small drops than large ones, or that as the intensity of a shower increases (i.e. with heavier rainfall) the number of large drops relative to small ones grows. Nor would they have known the range of sizes of raindrops. The smallest are a mere 0.2 mm in diameter, equivalent to the thickness of two or three pages of this book; drops smaller than this are not heavy enough to fall all the way to the ground under their

---

[9] Newton never achieved a well separated spectrum because his source of light was a small circular hole of diameter of ¼ inch (6 mm). It was William Wollaston who in 1802 first created a spectrum using a very narrow slit, an arrangement that results in well separated colours in which dark absorption lines were visible.

[10] Descartes, R. (2001), p 319.

own weight before evaporating entirely. At the other end of the scale, the largest raindrops can sometimes reach a diameter of 6 mm, though only in the most intense tropical downpours. Moreover, very large drops are buffeted by the air through which they fall, which constantly alters their shape, and so can't contribute light to the arc of a rainbow. Drops larger than this are not possible because as they fall they would break up into smaller ones due to air turbulence.

Both men realised that their respective accounts of the rainbow depend on raindrops being perfectly spherical, for, as Descartes pointed out, "they may not lose the smallest part of their roundness without this making a notable difference in the angle under which the colours must appear."[11] But neither of them ever considered that the size of a raindrop has an effect on the rainbow. Indeed, Descartes went so far as to say that drops "being larger or smaller does not change the appearance of the arc"[12] Newton agreed: "Since now raindrops are extremely small with respect to their distance from an observer's eye, so that physically they can be considered points, we need not at all consider their size."[13] How wrong they were. As we shall see in the next chapter, drop size is the single most important factor that determines both the size of a rainbow and its colours.

In fact, the effect of drop size on the colours in a rainbow had been noticed a century and half earlier by that eagle-eyed polymath, Leonardo da Vinci (1452–1519). In his explanation for the redness of rainbows formed when the sun is on the horizon he adds "That redness, together with the other colours [of the rainbow], is of much greater intensity, the more the rain is composed of large drops, and the more minute the drops are, the paler are the colours. If the rain is of the nature of a mist, then the rainbow will be white and so completely without hue."[14]

Nor are all raindrops perfectly spherical. Air friction alters the shape of any drop that has a diameter of more than 1 mm as it falls through the air, reducing the curvature of the lower half. The distortion increases with drop size so that the largest stable drops resemble miniature buns. The change in shape has a marked effect on the appearance of a rainbow because the path of light through a drop that is not perfectly spherical is not the same in every plane. Light is deviated to a greater degree as it traverses the vertical cross-section of an oblate drop than it does as it traverses the horizontal one

---

[11] Descartes, R. (2001), p 343.

[12] Descartes, R. (2001), p 332.

[13] Newton, I. (1984) p 595.

[14] Leonardo da Vinci (1956), p 335.

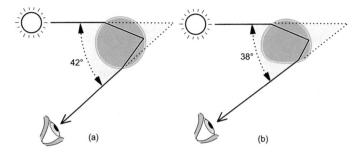

Fig. 7.1 The path of the rainbow ray **a** through a spherical drop and **b** through a flattened one. Compared to the path of the rainbow ray through a spherical drop, the rainbow ray through a flattened drop is deviated to a greater degree from the direction it enters the drop. The net result is a rainbow with a smaller radius

because its vertical cross-section is approximately elliptical while its horizontal one remains more nearly circular.

The first person to investigate how oblate drops might affect the shape and colours of a rainbow was Giambatista Venturi . Venturi demonstrated both mathematically and experimentally that rainbow rays from oblate drops are deviated more than those from perfectly spherical drops.[15] Hence the presence of oblate drops in a shower of rain should result in a rainbow arc just inside the apex of a rainbow formed in spherical drops (Fig. 7.1).[16]

Rainbow rays from oblate drops near the ground, however, emerge in the same direction as those from smaller drops because the horizontal cross-section of all drops, large and small, is circular. Thus drops of all sizes contribute to the colour and brightness of the foot of a rainbow, which is why is why the foot of a rainbow seen in a heavy downpour is often brighter and its colours noticeably more vivid than at its apex, though this can occur only when the sun is at the horizon. Individually, however, a small drop reflects less light than a large one, so all else being equal, a rainbow due to small drops is intrinsically less bright than one due to larger drops.

The increased brightness near the base of a rainbow formed by large drops in heavy rainfall may well have lent credence to the folk tale that there is a pot of gold at the foot of a rainbow, though, as we saw in chapter three, in itself is unlikely to have been the origin of that belief. Given the rainbow's supernatural associations, the ground it touches would have had a special significance for pre-scientific folk, however bright the foot of the bow.

Following the publication of Newton's *Opticks* at the beginning of the eighteenth century, it seemed as if the geometrical explanation of the rainbow

---

[15] Venturi, G., B, (1814). See: Tav. VIII, Fig. 13, for the figure of the apparatus he used.
[16] Venturi, G., B. (1814), pp 166–80.

left no stone unturned. Yet its very success sowed the seeds of its demise because it encouraged people to look at rainbows more attentively. Now that they knew what to look for, people began to take a greater interest in unusual rainbows that had hitherto been overlooked or whose significance had been ignored. And in increasing numbers they submitted detailed descriptions of their observations to the newly established scientific journals of the day such as the *Philosophical Transactions of the Royal Society* and the *Memoires of the Académie Royale des Sciences*.

At the time it seemed that the Newtonian account could be stretched to accommodate these new observations. Halley used it to explain the reflection bow that he had seen in Chester. He realised that in the right circumstances, the sun's reflection provides a second source of light that is, in effect, as far below the horizon as the sun is above it and which gives rise to a second primary bow that encircles the primary bow formed by direct sunlight. The higher the sun is in the sky, the more of the arc of this second primary is visible and the less that of the direct primary. The arc of this second primary, however, can be formed only if the body of water in which the sun is reflected is large—as large as the bow to which it gives rise. You won't see a reflection bow if the reflection is from a small pond. And, as one would expect, the bow formed by reflected sunlight is markedly less bright than the one formed by direct sunlight. Furthermore, depending on how choppy the water is, the colours in the reflected light primary bow may either be blurred or absent in parts of the arc because each wave reflects sunlight in a different direction to its neighbours so that the rain is not illuminated by light reflected entirely in the same direction.

Although a rainbow can be formed by reflected light, it isn't possible to see the reflection of a rainbow itself because, as Descartes had established, all the rays responsible for a rainbow must enter the eye directly from the drops in which it is seen (Fig. 7.2). That, of course, is why we can't see the same rainbow as one another. But in the right circumstances a rainbow over open water can be accompanied by what appears to be its reflection, something that, on the face of it, should be impossible. The reflection is actually that of a rainbow formed by a different set of drops to those in which you see the bow in the sky. It is a reflection of an 'invisible' rainbow, one that would be seen by someone whose eye was as far below the horizon as yours is above it, i.e. by your mirror image.

A tell-tale sign that the reflection is not that of the rainbow in the sky is that the ends of the two bows meet at the horizon rather than at the foot main rainbow, as the following description makes amply clear.

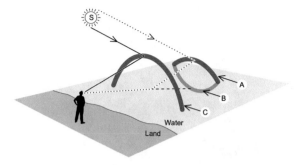

Fig. 7.2 The geometry of the reflection of a rainbow. The reflection, B, seen on the surface of the water is of a bow, A, that can't be seen directly by the observer. C is the rainbow seen in the sky

> A few weeks ago I had the pleasure of seeing a rainbow and its reflection, or at least a reflection of one from the same shower at the same time, in smooth water. The base of the bow in the cloud seemed but a few hundred yards from me, and the reflection evidently did not belong to it as the two bases did not correspond, the reflected bow lying inside the other, the red of the one commencing where the violet rays of the other disappeared (Fig. 7.3).[17]

The limitations of Newton's geometrical rainbow became increasingly apparent when observers began to demand an explanation for the supernumerary arcs that are sometimes seen just inside the uppermost section of the primary bow. Remarkably, these went largely unacknowledged until well into the eighteenth century. Although they don't appear every time a rainbow is seen, they are hardly rare. Yet, such is the hold that our preconceptions can have over us that they were seldom remarked upon in previous centuries. One of the few scholars to have noted their presence was Theodoric. Neither Descartes nor Newton, a particularly sharp-eyed observer, appears to have known of them. And it was Theodoric's description of arcs "immediately contiguous to the lower [i.e. primary] iris" that led Venturi to investigate the effect of oblate drops on the rainbow in the belief that oblate drops are responsible for supernumerary arcs.[18]

One of the earliest reports of these elusive arcs to be widely circulated appeared in the Philosophical Transactions in 1722.

> When the primary rainbow has been very vivid, I have observed in it, more than once, a second series of colours within, contiguous to the first but

[17] Dawson, G. (1874), p 322.
[18] Venturi, G B (1814), p 166.

Fig. 7.3 Rainbow reflected in the still waters of the Coorong National Park, South Australia[19]

far weaker, and sometimes a faint appearance of a third. These increase the rainbow to a breadth much exceeding what has hitherto been determined by calculation.[20]

"Determined by calculation" was a reference to Newton's geometrical account of the rainbow. The author, Benjamin Langwith (1684–1743), a clergyman, went on to describe in great detail supernumerary arcs that he had seen on four separate occasions. He noted that the colours in these bows were not a repetition of those in the main arc but alternating bands of purple and green that grew narrower and fainter with every repetition. He concluded.

I begin now to imagine, that the Rainbow seldom appears very lively without something of this Nature and that the supposed exact Agreement between the colours of the Rainbow and those of the Prism, is the reason that it has been so little observed.[21]

[19] Photo by Mundoo, Wikimedia Commons. https://tinyurl.com/ptd9rxhp Accessed 1/10/22.
[20] Langwith, B. (1723), p 241.
[21] Langwith, B. (1723), p 243.

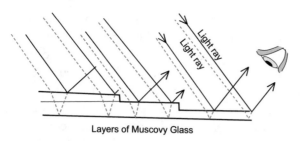

Layers of Muscovy Glass

Fig. 7.4 Simplified and annotated copy of Hooke's explanation of the origin of colours seen in *Muscovy Glass* (mica). Solid lines represent light reflected from the surface of the mica. Dotted lines represent the path of light that refracts into the mica and is reflected from the lower surface. The length of the path of light through the mica depends on thickness of the layers. Colour is due to the recombination both reflections

Langwith's letter was shown to Henry Pemberton (1694–1771), a professor of medicine at Oxford University and an able mathematician, chosen by Newton to edit the third edition of his mighty opus, the *Principia Mathematica*. Pemberton was struck by the similarity between Langwith's arcs and Newton's descriptions of the colours seen in thin films and suggested that these might hold the clue to what later came to be known as supernumerary arcs.

As we saw in the last chapter, Newton became interested in this phenomenon as a result of reading Hooke's *Micrographia*.[22] Hooke had found that these colours depend on the thickness of the film in which they are seen. He drew attention to the "several consecutions of colours" that can be seen in a thin wedge or lamina of *Muscovy-glass* "whose order from the thin end towards the thick, shall be Yellow, Red, Purple, Blue, Green; Yellow, Red, Purple, Blue, Green; Yellow, Red, Purple, Blue, Green; Yellow, &c. and these so often repeated".[23] But the necessary measurements, he admitted, were beyond him (Fig. 7.4).

> One thing which seems of the greatest concern in this Hypothesis, is to determine the greatest or least thickness requisite for these effects, which, though I have not been wanting in attempting, yet so exceeding thin are these coloured Plates, and so imperfect our Microscope, that I have not been hitherto successful.[24]

---

[22] Hooke studied thin-film colours in different substances such as soap bubbles and the oxidised surface of metals. See: Hooke, R. (2003), p 51.

[23] Hooke, R. (2003), p 67.

[24] Hooke, R. (2003), p 67.

When Newton investigated the phenomenon, probably soon after reading *Micrographia* in 1666, he came up with a typically ingenious solution that avoided the need to measure the thickness of a film directly. He rested a convex lens on a flat glass plate and pressed them together, thus creating a thin layer of air between them.[25] Taking his apparatus outdoors into daylight and looking down at the lens from above he saw the same sequence of colours as had Hooke, but as series of narrow bright, coloured concentric rings centred on the point of contact between the lens and glass plate. Knowing the curvature of the lens, he was able to calculate the distance between the glass plate and the curved surface of the lens and thus the thickness of the film of air corresponding to each bright ring. When he repeated the experiment in a darkened room, illuminating the apparatus with sunlight refracted through a prism to obtain a succession of pure spectral colours, he saw a series of concentric rings of the selected colour separated by dark bands. He noticed that the radius of a ring due to blue light was always less than that of the corresponding ring due to red light. This explained the pattern of colours of the rings in broad sunlight: it is due to overlapping rings of different spectral colours (Fig. 7.5).

So thorough was Newton's investigation and so detailed his account of his experiments, that the phenomenon, which might justifiably have been named Hooke's laminal colours, has since become known as Newton's Rings. It was the similarity between the sequence of colours in these rings formed in broad sunlight and those seen by Langwith in supernumerary arcs that led Pemberton to surmise, correctly as it turned out, that they both might be due to the same cause.[26]

The most remarkable aspect of Newton's investigation of these eponymous "rings" was the unprecedented precision with which he determined the distance between the glass plate and surface of the lens on which it rested. This enabled him to discover that the air gap between successive bright fringes always increases by a fixed amount, which in the case of yellowish-orange light he found to be a mere three ten thousandth of a millimetre, about 100 times less than the diameter of a human hair. Yet so accurate were his measurements that, as we shall see in the next chapter, they were used more than a century later by Thomas Young to calculate the wavelength of light corresponding to each of the colours of the spectrum.

---

[25] Hooke performed a similar experiment using "…two small pieces of ground and polished Looking-glass-plate … and with your fore-fingers and thumbs press them very hard and close together, and you shall find … there appear several *Irises* or coloured lines, in the same manner almost as in the *Muscovy Glass*…". See: Hooke, R. (2003), p 50.

[26] Pemberton, H. (1723).

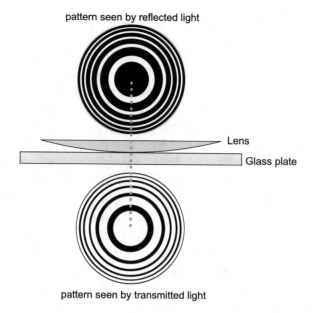

pattern seen by reflected light

Lens

Glass plate

pattern seen by transmitted light

Fig. 7.5 One way in which Newton's Rings can be studied. Newton used a double convex lens and pressed the flat side of a plano-convex lens against it

But how to explain these observations? Hooke's suggestion in *Micrographia* was that the colours come about because light reflected from the upper surface of the thin film combines with light reflected from its lower surface. He surmised that light reflected from the lower surface of the thin film is weakened due to refraction as it enters and leaves the film and "by reason of the time spent in passing and repassing between the two surfaces…"[27] When the reflection from the lower surface is combined in the eye of the observer with the reflection from the upper surface they produce colours because a weakened pulse is merged with a stronger one.

Newton, as we saw, had rejected Hooke's ideas on light and colour the moment he first came across them, principally because he believed that it is not possible to explain how light can travel in a straight line if it is a vibration of some sort. Yet to explain the regular pattern of repeated colours he grafted features of Hooke's wave theory onto his own particle theory and suggested that "assuming the rays of light to be small bodies … when they impinge on any refracting or reflecting superficies, must as necessarily excite vibrations … as stones do in water when thrown into it."[28] These vibrations, he suggested, might influence the motion of a particle causing it either to be refracted or

---

[27] Hooke, R. (2003), p 66.
[28] Newton, I., (1675 a), p 248.

reflected within a thin film due to what he called "fits of easy transmission or reflection".

He proposed that where the vibration set up in the æther is in the same direction as the motion of a particle, the particle is swept forward in a "fit of easy transmission", like a surfer catching a wave. And where the vibration opposes a particle's motion, the particle is forced back in a "fit of easy reflection". Looking through the lens at the glass plate, said Newton, the coloured rings are due to light that is reflected back into the eye from the air gap while the dark spaces between them are due to light that passes from the air to the glass plate on which the lens rests. Ascribing these "fits" to the effect of vibrations on the motion of particles had another advantage: if the alternation between bright and dark rings is due to the to-and-fro motion of a vibration, this might explain why their size increases by regular intervals.

But it seems that Newton later lost faith in his explanation for when he repeated it in his *Opticks* he concluded "But whether this hypothesis be true or false I do not here consider."[29] In fact, in the scathing phrase apparently once used by Wolfgang Pauli (1900–58), the Nobel prize-winning Austrian physicist, to dismiss a colleague's scientific paper, Newton's hypothesis was so far wide of the mark that "Not only is it not right, it's not even wrong". We now know that Hooke was much closer to the truth: the colours seen in thin, transparent films such as soap bubbles, oil slicks etc. are due to reflections that are minutely out of step with one another combining in the eye (or the optical instrument) of the observer.

Newton described his research into the "phænomena of thin plates" in the long letter to the Royal Society in 1675, mentioned in the last chapter. And, just as he had when asked to review Newton's prism experiments in 1671, Hooke again denied that Newton's findings were anything new, this time claiming that it was all to be found in *Micrographia*. Newton reacted angrily, pointing out that he had investigated the "phænomena of thin plates" far more thoroughly than had Hooke, who had "given no further insight into it than this, that the colour depended on some certain thickness of the plate; though what that thickness was at every colour, he confesses in his *Micrographia*, he had attempted in vain to learn; and therefore, seeing I was left to measure it myself, I suppose he will allow me to make use of what I took the pains to find out. And this I hope may vindicate me from what Mr. Hooke has been pleased to charge me with."[30]

---

[29] Newton, I. (1952), p 280.

[30] Newton, I. (1675 b).

Despite the rancorous exchange, a year later both men made a half-hearted attempt to patch things up. In 1676 N wrote to Hooke, saying

What Des-cartes did [in Opticks] was a good step. You have added much several ways, & especially in taking ye colours of thin plates into philosophical consideration. If I have seen further it is by standing on ye sholders of Giants.[31]

Given their frequent ill-tempered disagreements, the compliment may not have been entirely sincere; it was nonetheless true. Newton had learned a great deal about light from Hooke. But relations between them broke down irrevocably in 1686, when Newton was working on *Principia*. Hooke claimed, with some justification, that Newton owed to him the insight that the elliptical orbit of a planet is due to the gravitational pull of the sun on a body that would otherwise move in a straight line.[32] He demanded that Newton acknowledge the debt in *Principia*. Newton reacted angrily and threatened to stop work on the final section of *Principia*. "Now is this not very fine?" he railed at Halley, "Mathematicians that find out, settle & do all the business must content themselves with being nothing but dry calculators & drudges & another that does nothing but pretend and grasp at all things must carry away all the invention as well as those that were to follow as those that went before."[33] Newton didn't dispute that Hooke's suggestion had proved useful but, he maintained, the real issue was that Hooke was not enough of a mathematician to exploit the idea fully. That, of course, was why Hooke, for all his inventiveness, scientific imagination and experimental skill, could never match Newton as a scientist. Halley had to exert all his diplomatic skills to pacify Newton and convince him of the importance of finishing the *Principia*. Predictably, when it was finally published in 1687, all references to Hooke in the original draft had been excised.

Later, with the publication of *Opticks*, Newton seized another opportunity to cut those giants down to size upon whose shoulders, as he had admitted to Hooke, he once had had to perch. Where the rainbow was concerned, Descartes was openly dismissed as a mere camp-follower to Dominis. As for Hooke's pioneering investigations of colour, they might as well never have taken place. Book II of *Opticks*, which deals with colours of thin transparent bodies, opens thus:

---

[31] Newton, I. (1676). In: Turnbull, W.H. (1959), p 416.
[32] Hooke had written to Newton about gravity in 1679. See: Turnbull, W.H. (1960), p. 297.
[33] Newton, I. (1686).

It has been observed by others, that transparent Substances, as Glass, Water, Air, &c., when made very thin by being blown into Bubbles, or otherwise formed into Plates, do exhibit various Colours according to their various thinness...[34]

No mention of Hooke; an unforgiving Newton had decided to settle old scores and expunge him from the history of optics.

Hooke died in 1703. In his final years he was plagued by the thought that his legacy would be eclipsed by Newton and ignored by his peers. His fears proved well founded and it is only in the latter half of the twentieth century that his reputation as one of the most important and ingenious natural philosophers of the seventeenth century has been restored.[35]

Someone else who had become interested in Hooke's account of colours in thin films at the same time as Newton was Christiaan Huygens (1629–1695).[36] Huygens was the only one of Newton's contemporaries who rivalled him as a mathematician and natural philosopher, so it is perhaps not surprising that he hit upon the same solution to the problem of measuring the thickness of a thin film: a convex lens resting upon a flat glass plate. But lacking Newton's obsession with accuracy, he was unable to match the precision of Newton's measurements and abandoned the investigation before it bore fruit (Fig. 7.6).[37]

Huygens' background was very different to Newton's. He was born into one of the leading families of the seventeenth century Dutch Republic. His father, Constantijn Huygens (1596–1687), was a renowned Dutch poet and diplomat, a patron of Rembrandt and a close friend of Descartes. Christiaan received a first-class education at home and showed an early gift for mathematics and science and a Newtonian flair for experimentation, as well as a marked talent for music and drawing.[38] He went on to study law and mathematics at Leiden University where he attended Frans van Schooten's mathematics lectures. Frans van Schooten (1615–60) was a committed Cartesian and the editor of the 1649 Latin edition of Descartes' Geometry, the very edition from which Newton was to learn about modern mathematics as an undergraduate.[39]

---

[34] Newton, I. (1952), p 193.

[35] Chapman, A. (2004).

[36] For Huygens' life and scientific work see: Bell, A.E. (1948), Andriesse, C.D. (2005) and Aldersey-Williams, H. (2020).

[37] Huygens, C. (1932), pp 341–8.

[38] Huygens had a good singing voice and played the harpsichord.

[39] This edition included some of Huygens' early mathematical work carried out while he was Schooten's student.

"1665 November. Order of colours (from the center): *nothing, white, yellow, red, blue, green, yellow, blue, green*"

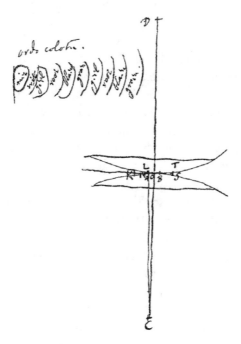

Fig. 7.6 Huygens' thin film experiment. The lower lens rests on a table covered with a dark cloth. Apparatus illuminated from above by skylight[40]

Christiaan was expected to become a diplomat, like his father, but much to the son's relief political events in Holland ruled out this career. In 1650 Constantijn's patron, the Prince of Orange and Stadholder of Holland, William II (1624–50), lost power, and with it went the Huygens' family political clout. And so, after completing his university studies, Christiaan returned home, where he stayed for the next ten years, applying his formidable intellect and inventiveness to mathematics, physics and astronomy while supported financially by his father. Together with his elder brother, Constantijn, he designed and constructed a number of astronomical telescopes from scratch, each an improvement on the last, which enabled him to make several important astronomical discoveries. On the night of 25th March, 1655 using a telescope capable of magnifying 50 times he discovered Saturn's largest moon,

---

[40] Huygens, C. (1932), pp 341.

which we now know as Titan.[41] And over several months during the winter of 1655/56, using another of his telescopes that could magnify 100 times, he patiently unravelled the puzzle of Saturn's ring system, which every other astronomer, beginning with Galileo, with their inferior telescopes, had been unable to see clearly and had mistaken for accompanying stars.[42] By March 1656 Huygens had established that "[Saturn] is encircled by a ring, thin, plane nowhere attached, inclined to the ecliptic".[43] Three years later, in November, 1559, while observing Mars, he made the first ever drawing of a surface feature on another planet. His sketch shows a large Y-shaped patch, now known as Syrtis Major, in the middle of the planet's the disc.[44]

Without any doubt, however, his greatest achievement during those years was to construct the world's first accurate clock, the design of which was based on his mathematical investigation of the motion of the pendulum. Huygens hoped that it would provide the solution to one of the most pressing problems of the age: the accurate determination of longitude at sea. But sea trials showed that as it ploughed is way through choppy seas, a ship's uneven movements interfered with the regular swing of the clock's pendulum, which made his design impractical for use aboard ships.[45] Huygens eventually got around to publishing a full description of the clock and the mathematical principles underlying its construction in 1673. The title of the book was *Horologium Oscillatorium*, and it would have been among the most important works on the science of motion of the seventeenth century had it not been for Newton's *Principia*.[46]

By 1660 his discoveries and inventions had established his reputation as Europe's leading natural philosopher and one of its most able mathematicians. He spent some time in Paris and attended meetings of its leading savants. The following year he paid a brief visit to London to learn at first-hand about a newly formed scientific society that was to become the Royal Society a year later. Although he considered that London as a city compared

---

[41] This was the first discovery of a moon orbiting a planet since Galileo's discovery of Jupiter's four moons in 1610. Curiously, Huygens did not give it a name. That was done by John Herschel (1792–1871) in 1847 when he proposed that the moons of Saturn that were known at the time should be named as *Mimas, Enceladus, Tethys, Dione, Rhea, Titan,* and *Iapetus*. There is a certain irony in naming the 1997 Saturn probe *Cassini* and the Titan lander *Huygens* because the two men did not get on; Cassini was jealous of Huygens' reputation and Huygens resented the fact that Cassini was better paid than he was.

[42] Shapley, D. (1949).

[43] Huygens, C. (1659). See: Bell, A.E. (1948), pp 30–34.

[44] Sheehan, W. (1988), p 42.

[45] For a brief account of how the longitude problem at sea was solved using clocks see: Sobel, D. (1998).

[46] Huygens, C. (1673).

unfavourably to Paris—he said that it was dirty, smelly and cramped—he found that the meetings of the English natural philosophers far more interesting and productive than those of the amateur scientific societies in Paris.

He was made a Fellow of the Royal Society in 1663 and a year later was invited to live and work in Paris by Jean Baptiste Colbert. By attracting the best brains in Europe to Paris, Colbert planned to make France the leading cultural power in Europe and a land worthy of its king. But it was only when he was persuaded by the leading Parisian savants of the advantages of organising their collective efforts for the good of the nation that Colbert agreed to the creation of the Académie Royale des Sciences, the French equivalent of the Royal Society.[47] Unlike the Royal Society, however, the Académie was a state-funded institution and its membership restricted to savants judged to be of use to the nation.[48] Hard-line Cartesians were excluded because Colbert considered that they were excessively dogmatic. Jesuits were also barred because they were reckoned not to be sufficiently open-minded.

Once the Académie was established in 1666, Huygens was made its director and in no time became its leading light.[49] But although he felt very much at home in France, as a Protestant in a Catholic country, he had to rely on Colbert's patronage to keep potential enemies at bay. Indeed, when Louis XIV, invaded the Low Countries in 1672, and every Dutchman was ordered to leave France, Colbert assured Huygens that the order did not apply to him. Nor did Huygens feel under any obligation to join the exodus, claiming "there is nothing in the post that I occupy that has anything to do the with war."[50]

But if loyalty to his country could not drive him from France, frail health often did. Throughout his life Huygens suffered prolonged periods of crippling depression, sometimes taking a year or more to recover. During his long sojourn in France he returned home to Holland on sick leave on three occasions, the last being in 1681, two years before Colbert's death. It was while he was recovering from one of these depressions in 1677 that he began developing his ideas on light, which he completed and wrote up as a manuscript

---

[47] George, A. J. (1938), p 382.

[48] During Colbert's life there were never more than 20 Academicians. See: Stroup, A. (1987), pp. 14–15.

[49] Huygens received 6000 livres p.a., while French savants received an average of 1,500 livres apiece. See Briggs, R. (1991), p 62.

[50] Huygens, C. (1895), p 544.

titled *Traité de la Lumière*[51] when he returned to France in 1678. However, he didn't get around to publishing it as a book until 1690.[52]

The long delay between the completion of the *Traité* and its publication, as was the case with his *Horologium Oscillatorium,* was typical of Huygens, and goes some way to explaining why he never exerted an influence commensurate with his ground-breaking inventions and discoveries. He was something of a one-man band, preferring to work on problems that interested him alone and was often reluctant to publish or even to publicise his work. Yet he was the very model of a modern mathematical physicist, arguably the very first of that breed, usually favouring mathematical modelling over experiment as a way of getting to the heart of a problem. But he was not a systems-builder in the mould of Descartes, or indeed Newton. He was uninterested in the wider picture at a time when the educated public looked to its leading natural philosophers for a compressive account of the world at large, or, as we might say today, a "theory of everything".

In 1681 he had planned to return to Paris as soon as he was well enough, but learned from friends in Paris that he was no longer welcome there. Not only were some of his erstwhile colleagues in the Académie opposed to his return, Christiaan's elder brother, Constantijn, was now secretary to William III (1650–1702), Prince of Orange, and Louis' bitterest enemy.[53] At the same time, France was fast becoming unsafe for Protestants—this was only four years before the revocation of the Edict of Nantes by Louis XIV, which made Protestantism illegal in his kingdom. Moreover, Louis had the Dutch in his sights once again, having waged a successful war against them only a few years earlier,[54] and was busy planning another invasion of Holland as part of his grand plan to make France Europe's dominant political power.

Huygens visited England once again for a few months in 1689, during which he finally met Newton "whom I exceedingly admire for the beautiful inventions that I found in the work [*Principia*] that he sent me."[55] The heated exchange of 1673 over the nature of light and colour had long since been forgotten. Indeed, he tried to use his family connections, unsuccessfully as it turned out, with the newly established William III, to help Newton's bid to be appointed Master of Trinity College.

---

[51] Huygens, C. (1690).

[52] The *Traité de la Lumière* was written in French. It was translated into English: Thompson, S.P., 1912, Treatise on Light. Surprisingly, there was no Dutch edition of this seminal work until 1990: "Christiaan Huygens: Verhandeling over het licht". (trans. Eringa, D.).

[53] William III was the only son of Willhelm II.

[54] The Franco-Dutch War 1672–78.

[55] Huygens, C. (1901), p 305.

During his last years, which he spent in Holland, he felt increasingly isolated and worn out. His health broke down again in 1694, but this time there was to be no recovery. He lost weight and fell into a deep depression. Six weeks before his death in July, 1695, his brother, Constantijn, rushed to his side.

I arrive to find brother Christiaan sorely ill. He complains of pain and of bedsores. Anything with which he may harm himself has been removed. He began to cut himself with broken glass and prick himself with pins. Also he stuck a marble down his throat. His servant heard it rattle and managed to extricate it by slapping upon his back. He cried out then, 'Slap hard'. Sometimes [he] dreams and hears people speak who are not there. Says that people would tear him apart if they heard his view on religion. Hopes that he will not be held to blame for this view, for he is out of his senses. Sometimes screams loudly and curses.[56]

Christiaan's family always maintained that his depressions were in part brought on by his lack of religious faith. The day before he died he reluctantly agreed that a priest should be called to his bedside, though no one was persuaded that this was a sincere last minute attempt to make his peace with his Maker.

Although Newton and Huygens had come up with a similar solution to the problem of measuring thin films that had defeated Hooke, on the question of the nature of light they were poles apart. Despite his explanation of the colours in thin films in terms of vibrations that brought about "fits of easy transmission and reflection", Newton's abiding fixation with rectilinear propagation meant that he could never wholeheartedly set aside his belief that light is essentially corpuscular because, he maintained, only particles can move in straight lines, whereas waves always spread out in every direction. Huygens disagreed: he pointed out that two beams of light can cross without affecting one another, something that waves can do but which would be impossible if light were a stream of Newtonian corpuscles. Instead, he developed a mathematical version of Hooke's wave theory, major aspects of which have endured to this day.[57]

Despite being highly critical of Descartes' wildly speculative explanations of natural phenomena—recall his disparaging comments on Descartes' account of the origin of colour—Huygens nevertheless subscribed to the twin Cartesian principles that all natural events are the result of direct contact

---

[56] Andriesse, C.D. (2005), p 410.
[57] Barth, M. (1995), pp 601–13.

between material bodies and that a scientific explanation must reveal the mechanism responsible for a phenomenon: "...in the true Philosophy," he wrote, "...one conceives the causes of all natural effects in terms of mechanical motions. This, in my opinion, we must necessarily do, or else renounce all hopes of ever comprehending anything in physics."[58] But if Huygens was a Cartesian, it was in the spirit of the Descartes who had explained the rainbow mathematically rather than the Descartes who had come up with ad hoc hypotheses about the nature of light and colour.[59]

In contrast to Hooke, Huygens assumed that light is transmitted through the æther as a succession of solitary pulses. Unlike Descartes' æther, which consists of matter that is completely rigid and inelastic, Huygens' æther is composed of equally sized, highly elastic particles that are tightly packed together. He had proved mathematically several years earlier that under these conditions an impulse due to a collision will pass from one particle to the next without the particles themselves moving, just as happens in the scientific toy known today as a Newton's Cradle. Indeed, Huygens had constructed a Newton's Cradle using glass spheres when he was developing his ideas on collisions between bodies. Crucially, the speed at which an impulse travels through a collection of such particles, although large, is finite, whereas in Descartes' absolutely rigid æther every disturbance would propagate instantaneously to the furthest reaches of the universe. And in another important departure from Descartes, Huygens assumed that the speed of an impulse is less in glass and water than it is in air or a vacuum.

Huygens based his conjecture that the speed of light is finite on his theory of collisions, but he was delighted to have this confirmed in 1676 by a discovery made by Ole Rømer (1644–1710), one of his Parisian colleagues.[60] Rømer was a Danish astronomer recruited by Colbert, as Huygens had been, to the Académie Royale. Between 1671 and 1673, using one of Huygens' pendulum clocks, which allowed astronomers to measure intervals of time with hitherto unprecedented accuracy, Rømer had made a series of systematic observations to compile a table of the times of eclipses by Jupiter of its innermost moon, Io.[61] The purpose was to use the eclipses as a universal

---

[58] Huygens, C. (1912), p 3.

[59] It could be said that the Cartesian faith in 'mechanical motions' survived well into the twentieth century, eventually falling foul of the positivism of quantum theory.

[60] Huygens, C. (1899), pp 36–7.

[61] The best of Huygens' clocks were accurate to 10–15 s per day, a 60-fold improvement on previous clocks.

clock that would enable clocks in other parts of the world to be synchronised with one at known longitude in order to determine one's longitude.[62]

To Rømer's surprise, he found that as the earth in its orbit about the sun moves away from Jupiter, the interval between successive pairs of eclipses of Io increases very slightly. The opposite happens as the Earth moves towards Jupiter: intervals between successive eclipses diminish slightly.[63] He realised that this was direct evidence of the finite speed of light for if, as Descartes had claimed, light travels instantaneously, then the interval between successive eclipses would be unaffected by the Earth's distance from Jupiter. From his observations Rømer calculated that, observed from the earth, the moment that Io is eclipsed by Jupiter when the distance between the two planets is greatest (i.e. when they are on opposite sides of the sun, known in astronomy as a conjunction) occurs twenty-two minutes later than expected compared to the one when the planets are at their closest (i.e. when they are on the same side of the sun, known as opposition). In other words, it takes light twenty-two minutes to cover a distance equal to the diameter of the earth's orbit. An account of his discovery was published in 1676.[64]

But it was Huygens rather than Rømer who used these observations to calculate the speed of light—Rømer's only word on the subject was "that to traverse a distance of about 3000 leagues, which is almost the diameter of the earth, light does not need a second of time.", i.e. light travels very fast.[65] Serendipitously, during the years that Rømer was timing the eclipses of Io, another of Colbert's imports, the Italian astronomer Domenico Cassini (1625–1712) had established a value for the distance between the earth and sun that is within 2% of that accepted today. Together with Rømer's estimate of the time it takes light to traverse the diameter of Earth's orbit, Huygens calculated that the speed of light is (in modern units) 221,000 km/s, which considering that this was the first estimate of the speed of light and given the limitations of the instruments and methods available to the astronomers of

---

[62] The idea that eclipses of Io could be used to determine longitude had originally been proposed by Galileo a couple of years after his discovery of Jupiter's moons in 1610, but he never got around to making the necessary observations. Rømer's observations were made from the Paris Observatory, where he lodged during his stay in France.

[63] The interval between successive eclipses varies, but is never more than about half a second. The difference becomes obvious only over several eclipse cycles.

[64] Rømer, O. (1676).

[65] Rømer, O. (1676), pp 233–4. English translation in: Magie, W.M. (1963), pp 335–7.

Fig. 7.7 Huygens' drawing of pulses of light emitted by a candle flame due to the motion of particles A, B and C[68]

the day is remarkably close to the value accepted today.[66] Huygens included the calculation in his *Traité de la Lumiere*.[67]

As to the nature of light, Huygens' assumed that fragments of matter within a source of light move about randomly at high speed and that a pulse is created in the surrounding æther every time it is struck by one of these fragments, giving rise to innumerable uncoordinated pulses within it (Fig. 7.7). To picture what happens next, he said, imagine dropping a stone into a pool and watching the resulting ripple travel across its surface. The ripple moves away from the point of impact as a circle of ever-increasing diameter. To determine the progress of the ripple, Huygens postulated that every point on its circumference can be considered to be a source of a miniature ripple, which he called a secondary wave. He could then calculate where the ripple as a whole would be a short while later by finding where all the secondary waves are in step with one another at that moment. In this way, the progress of a wave can be precisely tracked over time. The technique is known as Huygens Principle, and is still employed today to show how reflection, refraction and diffraction bring about a change in direction in a beam of light.

---

[66] Rømer overestimated the maximum delay between eclipses. It is 16 min 36 s, not 22 min. The accepted value of speed of light is 299,792.5 km/sec. The error is almost certainly a result in measuring the time at which the eclipses occur, Hughens' clock notwithstanding. See: Shea, J.H. (1998).

[67] Huygens, C. (1912), pp 7–10.

[68] Huygens, C. (1912), p 17.

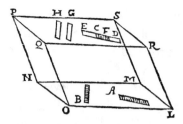

Fig. 7.8 Bartholin's drawing showing the effects of double refraction in Iceland spar. When the marks A & B are seen through the crystal, B splits into H & G. A produces two overlapping images, EF & CD[70]

Huygens employed the technique to derive the already known laws of reflection and refraction of light. But it proved particularly fruitful when he used it to explain a curious optical phenomenon that had first been noticed by Erasmus Bartholin (1625–1698), a Danish medical man and mathematician and, coincidentally, Rømer's father-in-law and former teacher. Bartholin had been given some specimens of a transparent crystal that had recently been discovered in Iceland at the site of an extinct volcano and noticed: "…a wonderful and extraordinary phenomenon: objects which are looked at through the crystal do not show, as in the case of other transparent bodies, a single refracted image, but they appear double."[69] Even more intriguingly, when the crystal is rotated, one of the images remains stationary while the other one moves around it in a small circle. He ascribed the stationary image to ordinary (i.e. Cartesian) refraction and the one that moved to what he called extraordinary refraction and published his research as small pamphlet in 1669. Today the phenomenon is known as *double refraction* or *birefringence* (Fig. 7.8).

A couple of years later, in 1672, Huygens was given a large piece of Cristal d'Island, or Iceland spar as it is commonly known in English, by Jean Picard (1660–1682), a fellow Academician who had acquired it while in Denmark where he had gone to make some astronomical observations to establish the longitude of Uraniborg, Tycho Brahe's old observatory.[71] Huygens carried out his own investigation into its curious properties with a thoroughness and precision to rival Newton at his best. He eventually realised that he could account for ordinary and extraordinary refraction by assuming that light travels through Iceland spar with two velocities, one of which—that of

---

[69] Bartholini, E. (1669). English translation in: Magie, W.M. (1963), pp 280–3.

[70] Bartholini, E. (1669), p 12.

[71] Iceland Spar wasn't the only thing that Picard brought back with him, he also brought Ole Rømer. See: Ziggelaar, A. (1980), pp 181–2.

the extraordinary ray—is determined by the arrangement of the atoms within the crystal. Huygens was so pleased with his explanation of double refraction that he declared that it provided the "...*experimentum crucis that* confirms my theory of light and refraction."[72]

He had spoken too soon, for when he viewed the double refraction due to one crystal through a second crystal, depending on their mutual orientation, the two refractions either continued through the second crystal unaffected or else swapped places: the stationary image due to ordinary refraction was transformed into the mobile image due to extraordinary refraction and vice versa. Huygens admitted defeat at this point, adding with admirable candour and generosity, "For though I have not been able till now to find its cause, I do not for that reason wish to desist from describing it, in order to give opportunity to others to investigate it."[73]

Newton seized upon this admission as further evidence that light is not a wave. He suggested that Huygens' observation could be explained by supposing that light has "some kind of attractive virtue lodged in certain Sides both of the rays, and of the particles of the Crystal",[74] or, to use modern terminology, that light is itself polarised. Huygens had assumed that double refraction depends entirely upon what happens to light within Iceland spar, whereas Newton surmised that whatever property is revealed by its passage through this crystal, light possesses that property before it enters the crystal. For Newton, the crystal merely acts like a sieve, separating light polarised in one direction from that polarised in the perpendicular direction, whereas Huygens maintained—incorrectly as it happens—that polarisation is brought about within the crystal.

At the time, double refraction was considered to be an unusual, if interesting, phenomenon confined to Iceland spar that had no bearing on the wider question of the nature of light. It was, however, an epoch-making discovery to rival that of the heterogeneity of white light, though its full impact was not felt for almost another century and a half. As we shall see in the next chapter, double refraction offers a most important clue to the nature of light, a clue that Huygens missed and which Newton guessed, if imperfectly. Nevertheless, the honours in this contest must go to Huygens, for his mathematical analysis that enabled him to predict the path taken by the extraordinary ray was spot on, whereas Newton's was not.

Huygens was the only scientist in the seventeenth century to come up with a mathematically rigorous account of the laws of reflection and refraction and

[72] Huygens, C. (1888), p 613.
[73] Huygens, C. (1912), p 92.
[74] Newton, I. (1952), p 373.

of double refraction based directly on ideas about the nature of light, and the contents of *Traité de la Lumière* represent the greatest theoretical advance in the understanding of the nature of light during the seventeenth century. But his explanation for the rectilinear propagation of light was unsatisfactory, and failed to convince his peers. Why doesn't a pulse spread out when it passes through an opening? Huygens' disappointing answer was that the secondary waves near an edge are too feeble to create a pulse within the shadow beyond an opening. As for colour, Huygens didn't even attempt an explanation. Nor could he, for his conception of light as a series of solitary pulses lacked the one property necessary to account for colour in terms of waves, namely periodicity, i.e. the repetition of pulses at regular intervals.

Newton, of course, had pointed out years earlier how such intervals might hold the clue to colour when he suggested that if Hooke's vibrations were "…of various depths or bignesses" then the size of these vibrations might determine the colour perceived by the eye. It is unlikely that Huygens would have been unaware of Newton's suggestion, but he never followed it up, possibly because he was unable to come up with a mechanism that would create a regular sequence of vibrations in the æther.

As we shall see in the next chapter, most of the properties of light can be explained by assuming that it behaves like a wave, not a series of solitary pulses as conceived by Huygens. A pulse is an isolated disturbance whereas a wave is a periodic one, one that repeats at regular intervals. And as Newton presciently suggested, if only to sneer at Hooke, these intervals hold the clue to colour.

And so the most promising account of the nature of light that the seventeenth century had to offer was first ignored and then forgotten, even before it was consigned to an undeserved oblivion by the enthusiastic reception of Newton's *Opticks*.

Long after he thought that he had finished with his optical investigations, Newton made a belated discovery that might have led him to revise his ideas about the nature of light: he found that light does not always travel in straight lines when he repeated—with his customary skill and thoroughness—experiments first made by a Jesuit priest, Francesco Maria Grimaldi (1618–1663).

Towards the end of a short but productive life—he was 45 when he died—Grimaldi, professor of mathematics at Bologna University and the man responsible for drawing one of the earliest maps of the moon, one that used a system of nomenclature on which all subsequent maps of the moon are based,[75] had carried out a wide ranging experimental investigation into the properties of light as they were then known, with a view to examining the Peripatetic, that is to say latter day Aristotelian idea that light is an accident, i.e. a quality, rather than a substance.

He wrote a lengthy book on the subject, *Physicomathesis de lumine, coloribus et iride* (*Physical and mathematical thesis on light, colour and the rainbow*), but died two years before its publication.[76] It is an admirably even-handed work, consisting of two lengthy sections, the first of which "introduces new experiments and reasoning that support the substantiality of light" while the second "refutes these arguments and teaches that it is probably possible to sustain the Peripatetic thesis of the accidentality of light."[77] But Grimaldi doesn't come down on one side or the other, so the reader with the stamina to plough through its 530 pages is left none the wiser, and the book might have joined the ranks of the scores of other long-forgotten seventeenth century optical texts had it not been for the series of experimental discoveries with which Grimaldi opens his treatise.

In a darkened chamber, he tells us, he had noticed that the shadow cast on a screen by a small object illuminated by a narrow beam of sunlight issuing from a very small hole high up in a widow shutter is considerably larger than expected—i.e., larger than it would be if light travels in straight lines and therefore larger than the geometrical shadow (Fig. 7.9a). Moreover, the shadow is always fringed with three faint multicoloured bands, in each of which the same colours arranged in the same order are visible: bluish on the side closest to the central shadow and reddish on the other side. In a variation of this experiment, he replaced the opaque object of the first experiment by a screen with a tiny hole in it. The resulting spot of light, falling on a white surface some distance beyond was, again, larger than expected and fringed with coloured bands of red and blue as in the first experiment (Fig. 7.9b). Even now that we know what to expect and with access to the latest apparatus these experiments are notoriously fiddly, so Grimaldi must have been a remarkably patient and skilled experimenter to have pulled them off, all the more so since he had no way of anticipating what they would reveal.[78]

Realising that he could not account for these observations in terms of either reflection or refraction, Grimaldi coined a new word for the phenomenon, *diffractio*, chosen because it suggests the spreading or breaking

[75] Montgomery, S. L. (1999), pp 198–208.

[76] Grimaldi, F.M. (1665), p 9.

[77] Ronchi, V. (1970), p 124.

[78] Grimaldi, F.M. (1665), p 1–11. See: Maggie, W.M. (1963), pp 294–8 for a translation of Grimaldi's observations of diffraction.

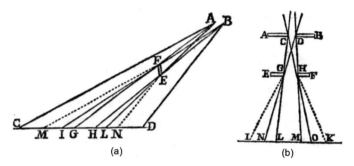

Fig. 7.9 a Grimaldi's interference. Sunlight enters a darkened room through a narrow opening AB. FE is a small disc casting an umbra GH and penumbra IL. Grimaldi noticed that the penumbra is surrounded by faint, coloured bands extending to M & I. **b** Grimaldi's diffraction. Sunlight enters a darkened room through a narrow hole CD. EF is a board, GF a hole larger than CD. The geometric edge of the spot of light should be NO. But the spot itself extends beyond this to I and K

up of light. Diffraction alone, however, does not explain the coloured fringes. These, we now know, are due to the interference of light waves, a phenomenon that was not properly studied until the start of the nineteenth century, and which is a subject for the next chapter.

Grimaldi suggested that his observations could be explained by assuming that light is a stream of some sort of fluid which, like water flowing around an obstacle, sets up eddies downstream from an object placed in its path. As anyone who has looked carefully at eddies formed in these conditions can testify, they are always wider than the obstacle, hence Grimaldi's assumption that something similar explained the diffraction of light. But he didn't pursue these speculations further. Nor did he suggest that light is vibration as Hooke was to do in *Micrographia*, published in the same year as Grimaldi's posthumous work.

Newton's knowledge of diffraction was for a long time second hand. He first came across Grimaldi in a book by another Jesuit, Honoré Fabri (1607–1688), who described the experiments on diffraction solely in order to discredit Grimaldi. Several years later, in 1675, Newton attended a lecture on "Several new Properties of Light, not observed, that he knew of, by Optick Writers" at the Royal Society in which Hooke described experiments he had performed involving diffraction.[79] There followed the, by now, inevitable clash between these two old adversaries: Newton disputed Hooke's claim that he had made new discoveries, saying that they were merely a new kind of refraction. Hooke replied indignantly "though it should be but a new kind of

---

[79] Hall, A.R. (1990), p 14.

refraction, yet it was a new one."[80] Once again, Hooke's intuition was keener than Newton's: it was a new phenomenon, though it had nothing to do with refraction.

At the time, Newton had, surprisingly, made no experiments on diffraction himself. Indeed, although the exact dates are unknown, he probably didn't carry out any such experiments before 1678, and may well have made his last ones as late as 1692, almost thirty years after he had completed the bulk of his research into optics and moved on to other things. Once he had the bit between his teeth, though, it was only a matter of time before his investigation of the phenomenon took him far beyond the point reached by either Grimaldi or Hooke. He devised imaginative experiments, made measurements with his usual astonishing precision, carefully recorded every detail of the procedures and observations, and eventually published them in the final section of *Opticks*.[81]

But the results of these experiments didn't shake his conviction that whatever else it might be, light cannot be a wave.. Rather than accept that diffraction is evidence that light is wavelike, he ascribed the spreading of light into the geometrical shadow and the accompanying coloured fringes to forces acting between the edge of the obstacle casting the shadow and the corpuscles of light that just grazed it. In some instances, he said, the force is attractive, in others repulsive. He even avoided using Grimaldi's terminology and renamed the phenomenon inflection. There speaks a man clutching at straws. Reading through the final pages of *Opticks*, it becomes apparent that he had at last reached the limits of his understanding concerning light. Diffraction is a phenomenon that he could not reconcile with his other views of light and which left his plan to provide a definitive account of the nature of light incomplete.

Although Newton completed a first draft of *Opticks* soon after his final series of experiments on diffraction in 1692, he continued to work on the manuscript intermittently until its publication in 1704. Pressed to explain why he was delaying publication, he replied it was "for fear that disputes and controversies may be raised against me by ignoramuses."[82] It may be that one of the ignoramuses he had in mind was Hooke. This can't be the whole story, however, for he was busy with other matters that kept him from his laboratory and away from his study. Although he remained Lucasian Professor until 1700, the year he was made Master of the Mint, he left Cambridge for

---

[80] Westfall, R.S. (1980), p 272.
[81] Newton, I. (1952), pp 317–406.
[82] Newton, I. (1693). In: Turnbull, W.H. (1961), 287; original in Latin, p 286.

London more or less permanently when he was appointed Warden of the Mint in 1696.

In those days this post was considered a sinecure, but at the time of Newton's appointment the currency of the realm was in crisis, its coinage having been seriously debased by forgery. Newton threw himself into the demanding job of overseeing the business of replacing all the coins in circulation throughout the kingdom, and proved admirably suited to the task due in no small part to his powerful mind and a methodical approach to every problem. All the while he continued to work intermittently on astronomical and mathematical problems and in 1703 was elected President of the Royal Society, the year of Hooke's death. A couple of years later, in 1705, he was knighted, not for his scientific achievements or for his work at the Mint but for services to the Crown.

But old habits die hard and he gave free rein to his testiness and paranoia on further occasions. In 1704 he fell out in a spectacular manner with the irascible John Flamsteed (1646–1719), the first Astronomer Royal, over the publication of Flamsteed's life's work, a new and accurate catalogue of the position of stars. Newton wanted it published despite Flamsteed's refusal to do so before it was completed, so that he could make use of its data to perfect his account of the moon's orbital motion for inclusion in the second edition of the *Principia*. Flamsteed, however, was no pushover and although Newton managed to have a bowdlerised version of the catalogue printed in 1712, Flamsteed, who knew how to push all Newton's buttons and delighted in doing so, prevailed in the end. He was allowed to collect and destroy the all the available unsold copies of the catalogue, while his own version—the *Historia Coelestis Britannica*—was eventually completed and published by his assistants six years after his death. Predictably, an enraged Newton set about removing all references to Flamsteed from the new edition of *Principia*, as he had done with Hooke's name in the first edition, but found to his chagrin that he couldn't do so entirely, having made extensive use of Flamsteed's observations in the section that dealt with the motion of comets.

Newton also clashed with one of Huygens' protégés, Gottfried Wilhelm Leibniz (1646–1716), the brilliant, hyperactive German lawyer, diplomat, philosopher, logician and, arguably, a greater mathematician than Newton, over priority concerning the invention of calculus. Newton had come up with a version of this important mathematical technique in 1666 when he was taking refuge from the plague in Woolsthorpe, while Leibniz had worked out its basic theorems independently during his time in Paris in 1675, and published his work soon after.[83] Newton, in the meantime, had kept his discovery to himself and was stirred to publish only after hearing that Leibniz had done so. Several years later, in 1712 Leibniz wrote a critical review of Newton's work in which he hinted that he, not Newton, was the true author of calculus. In the ensuing row, Leibniz was unwise enough to ask the Royal Society, of which Newton was then president, to adjudicate on the matter. Newton could not resist using his influence to outmanoeuvre Leibniz and a hand-picked committee of Fellows duly ruled in Newton's favour. But Leibniz had the last laugh, for the arcane mathematical notation Newton used in his calculus never found favour with the majority of European mathematicians, who preferred Leibniz's version. Indeed, it was mathematicians in Europe who, over the following decades, developed calculus into a powerful and indispensable scientific tool, while a misplaced allegiance to Newton left British mathematicians languishing unproductively in the doldrums for over a century.

Newton died in 1727 at his home in Kensington, then a village a few miles west of London, at the ripe old age of 85 from the effects of a blocked bladder, frail and with an occasionally unreliable memory but otherwise in full possession of his faculties, a full head of hair, good eyesight (it seems that he never had to use spectacles) and an almost complete set of teeth.[84] Faithful to the end to his heretical religious beliefs, he refused the last sacrament.

For the next two hundred years, Newton was considered to be the very embodiment of a rational approach to nature, and his interest in alchemical and religious matters was swept under the carpet. But with the rise of a new physics in the twentieth century, Newton no longer occupied centre stage, and it became possible to take a less partisan and more rounded view of the man and his legacy.[85]

John Maynard Keynes (1883–1946), the eminent economist, who bought a trunk-full of Newton's alchemical papers at an auction in 1936, wrote that:

> Newton was not the first of the age of reason. He was the last of the magicians, the last of the Babylonians and Sumerians, the last great mind which looked out on the visible and intellectual world with the same eyes as those who began to build our intellectual inheritance rather less than 10,000 years ago[86]

---

[83] Leibniz was in Paris on mission from his employer, the Elector of Mainz, to persuade Louis XIV to invade Egypt in the hope that it would deflect his ambitions in Europe. Louis didn't take up the idea, but a century later Napoleon did.

[84] According to a contemporary, Halley lost all his teeth in old age: "Dr Halley never eat any Thing but Fish, for he had no Teeth." In: Armitage, A. (1966), p 213.

[85] Fara, P. (2002).

[86] Keynes, J.M. (1947), p 27.

Keynes' conclusion has since been challenged as an anachronism. Newton, it has been pointed out, was a man of his time and a clear separation between science and religion, which we now take for granted, did not exist in the seventeenth century.[87] While that is true, when compared to many of his contemporaries, Newton does seem to belong to an earlier age. Neither Galileo, nor Descartes, nor Hooke, nor Huygens, nor Halley, to name but five of the leading natural philosophers of the seventeenth century, ever showed the least interest in the arcane subjects that so consumed Newton. They were recognizably men of a more secular, forward-looking era. Keynes' conclusion about Newton is probably not too far from the truth: he was never the icon of the age of reason for he had one foot in the past and believed that ancient wisdom was relevant to present concerns. Even though, as we shall now see, Newton became the very symbol of rationality of the eighteenth century Enlightenment, that era owed as much to men like Hooke and Huygens as it did to Newton. It is an open question whether Newton alone could have laid the foundations of the Enlightenment even if he had not been as interested in arcana as he was in natural philosophy. In any case, the Enlightenment was the product of many minds, philosophers, writers and historians among others.[88]

---

[87] Osler, M. J. (2006), p 302.
[88] Robertson, R. (2020).

8

Unweaving the Rainbow

Sir Isaac Newton, as we all know, was of the opinion that light was propelled from the sun, as a particle, in straight lines; Huygens and Hooke … supposed it to consist in a tremulous or undulatory motion; and there the matter rested. Cucumbers have continued to ripen, without waiting for the legal establishment of either hypothesis…[1]

The publication of *Opticks* in 1704 marked a belated close to the opening chapter of modern optics. The fact is that all the major properties of light—reflection, refraction, diffraction, interference and polarisation—had been discovered by 1675, thirty years before Newton finally got around to publishing the fruits of his optical researches. Of course, three of these—diffraction, interference and polarisation—were not recognised for what they are: the consequences of the wave nature of light. How could they be, given the undeveloped state of the wave theory at the time and, as we shall now see, that the rival corpuscular theory seemed to have all the answers? Far from inspiring fresh research into the nature of light, the publication of *Opticks* was followed by a long period of stagnation that lasted until the end the eighteenth century during which the rest of the world caught up with Newton, while largely ignoring Huygens and Hooke. Meanwhile, no doubt much to the relief of farmers and gardeners of a speculative turn of mind, plants

---

[1] Anonymous review of Thomas Young's wave theory of light. In: *The British Critic*, Vol XXV, January 1805, p 97.

© The Author(s), under exclusive license to Springer Nature Switzerland AG 2023
J. Naylor, *The Riddle of the Rainbow*, Copernicus Books,
https://doi.org/10.1007/978-3-031-23908-3_8

continued to thrive while the science of light languished for the best part of a century.

By the time of his death in 1727, Newton's ideas on light and gravity were known on the Continent, though they were generally viewed with scepticism, particularly in France. Although Descartes' star was by then on the wane, the Cartesian doctrine that a successful scientific explanation must always offer a mechanical explanation still held sway. Newtonian gravity, which supposedly kept planets in their respective orbits about the sun by means of a mutual attraction acting across empty space, was considered by most continental natural philosophers and mathematicians at best absurd and at worst an attempt to reintroduce occult powers into the natural world. Where they sought a mechanical explanation for gravity, Newton offered only his law of Universal Gravitation which was no more than a mathematical relationship between measurable quantities. Much to their consternation, Newton repeatedly insisted that he had no idea of the mechanism responsible for gravity: his final word on the subject in the penultimate paragraph of Principia was "Hypotheses non fingo"—"I do not feign hypotheses". But a few years later, in 1692, when pressed, he was prepared to explain why he had been reluctant to come up with a mechanism for gravity: "...that one body may act upon another at a distance through a vacuum without the mediation of any thing else by and through which their action or force may be conveyed from one to another is to me so great an absurdity that I believe no man who has in philosophical matters any competent faculty of thinking can ever fall into it."[2] How could this man be taken seriously as a physicist, asked Nicolas Malebranche (1638–1715), a leading Cartesian philosopher. That Newton was a great mathematician was indisputable, but by refusing to offer a mechanism to explain gravity he was surely no physicist![3]

As for light and colour, a generation earlier, Edmé Mariotte, then France's most accomplished experimentalist, had repeated the *experimentum crucis*, but unable to obtain pure colours as Newton had, concluded that Newton was mistaken about their cause.[4] Mariotte's failure to confirm the *experimentum crucis* led natural philosophers throughout Europe to turn their backs on Newton's ideas about light and colour until 1706, when the publication of a Latin translation of the *Opticks* gave them direct access to the wealth of his optical experiments and ideas.

But the man who probably did more than anyone to rescue Newton's scientific reputation in France and spread the Newtonian gospel abroad was

---

2 Newton, I. (1692/3) p 337.
3 Malebranche, N. (1978), pp 771–2.
4 Boyer, C. (1987), p 244.

neither a scientist nor an Englishman, he was an Anglophile man of letters, François Marie de Arouet (1694–1778), better known as Voltaire, his nom de plume, the author of mischievous satires such as *Candide*, which mocked Leibniz's philosophy of optimism, and of scurrilous political pamphlets that frequently led to trouble with the French establishment. In 1725, the fallout from one of these altercations forced him to take refuge in England for almost three years.[5]

During his enforced sojourn, Voltaire came to admire what he saw as the practical, down-to-earth nature of English philosophy and science, which he considered to be far superior to what he regarded as the French penchant for high-flown abstraction. Unlike the majority of French thinkers of the early eighteenth century, most of whom were in two minds about the relative merits of Descartes and Newton, Voltaire had no doubts: Descartes' fondness for ad hoc mechanical explanations, which all too often subordinated observation to plausible but mistaken hypotheses, was inferior to Newton's reliance on experiment and mathematics. Indeed, Voltaire became such a fervent convert to the Newtonian cause that when presented with a copy of Newton's Principia years later, he had a rush of blood to the head: "I have finally received the parcel sent by M. du Châtelet. It contains a copy of Newton. The first thing I did was to kneel down before it, as was only right."[6] His succinct description of the competing Newtonian and Cartesian world-views neatly illustrates the gulf between them:

> A Frenchman arriving in London finds things very different, in natural science as in everything else. He has left the world full, he finds it empty. In Paris they see the universe as composed of vortices of subtle matter, in London they see nothing of the kind. For us it is the pressure of the moon that causes the tides of the sea; for the English it is the sea that gravitates towards the moon, so that when you think that the moon should give us a high tide, these gentlemen think you should have a low one. … For your Cartesians everything is moved by an impulsion you don't really understand, for Mr Newton it is by gravitation, the cause of which is hardly better known. … For a Cartesian light exists in the air, for a Newtonian it comes from the sun in six and a half minutes.[7]

The reason why Newton was ignored in France, Voltaire concluded, was that few of his countrymen really understood Newton's natural philosophy. Hence

---

[5] Over the course of his life, Voltaire was twice locked up in the Bastille and exiled from France on three occasions.

[6] Davidson, I. (2010), p 132; see also Crossland, M. (1967), p 300.

[7] Voltaire (1980), p 68.

the need for a work that would make Newton's ideas on light and gravity intelligible to a wider French public. This project took him several years of intensive study and was published in 1738 as the *Élémens de la philosophie de Neuton*, a hefty tome of some 300 pages, to which he added a lengthy refutation of Cartesian science in a later edition.[8] It proved a great success. And since France was then the intellectual centre of Continental Europe, the favourable reception of Newton's ideas in France led to their gradual adoption by European savants. By the middle of the eighteenth century, Newtonian physics reigned supreme and Cartesian science was reduced to an historical footnote.

How a man of letters was able to write a successful account of some of the most complex and innovative scientific ideas of his time, even one devoid of mathematics, owed much to Emilie du Châtelet (1706–49), who was for many years his muse and lover. Voltaire made no secret of his debt to her: "Minerva dictates, I write", he declared. Mme du Châtelet was a gifted mathematician whose translation into French of the Principia, accompanied by her scholarly commentary of its finer points, and published a decade after Voltaire's *Élémens*, was considered to be superior to the original because it made Newtonian physics more accessible than it had been in Newton's Latin version.[9] However, she shared the misgivings of the majority of the Continent's thinkers about Newtonian gravity, so her direct contribution to the *Élémens* was probably limited to the section that deals with optics.[10]

Newtonian ideas about the nature of light, however, rested on foundations almost as shaky as those of Descartes, which meant that in optics the French swapped one dogma for another. Newton's ambivalence over whether light is particle or a wave was overlooked and the corpuscular theory became an unassailable scientific fact for the vast majority of French natural philosophers. Nevertheless, in the eighteenth century the wave theory had at least one very able advocate in Leonhard Euler (1707–1783), by common consent the most prolific mathematician in history and the most original scientist ever produced by Switzerland; he was also the first mathematician to apply Leibnitz's calculus to Newton's theories of motion. But Euler was almost alone in his belief that light can be explained in terms of waves. His distinctive contribution was the suggestion that atoms within a source of light vibrate with different frequencies and that they set up waves of the same frequency

---

[8] Voltaire (1738).

[9] Zinsser, J. P. (2001).

[10] Voltaire's account of Newton's explanation of the rainbow is in *Élémens*, pp 110–20.

The works of these men are still central to mainstream science: consult the index of any advanced textbook of physics and there you will find their names: Ampère, Arago, Biot, Fourier, Fresnel, Gay-Lussac, Lagrange, Laplace and Poisson.

in the surrounding æther, which, he postulated, is a continuous fluid, like water. Moreover, he suggested, the frequency of a wave determines the colour that is perceived. But Euler's ideas on light didn't catch on, in part because he didn't back them up with experiments, but principally because the world at large was in thrall to Newton.

By the end of the eighteenth century, the French had made themselves masters in the application of mathematics to physics and were considered, both at home and abroad, the true heirs to Newton. Indeed, as mathematical physicists, they were without peer.[11] Led by Pierre-Simon Laplace (1749–1827), an arch Newtonian and the author of the five volume *Traité de Méchanique Céleste*, a work that supplanted Newton's *Principia* as the vade mecum of physics, they sought to explain all natural events in terms of interactions between particles analysed mathematically using Newton's laws of motion and gravity.[12] Having long since abandoned the Cartesian views of their predecessors, French scientists now fully accepted the Newtonian assumption that bodies that are not in direct contact can exert forces directly upon one another without the need for an intervening medium. Moreover, they maintained, such action-at-a-distance occurs instantaneously.

One of Laplace's incidental achievements was to show that Newton's assumption that the Almighty has to intervene from time to time to ensure the integrity of the solar system was without foundation. Newton was convinced that the fact that the planets all orbit the sun in the same direction, and do so without colliding with one another, was evidence for God's existence. The clearest and most unequivocal statement of his views on this is to be found in his letters to Richard Bentley (1662–1742), an Anglican theologian.[13] In 1692 Bentley was preparing to give the first series of the Boyle Lectures, which had been endowed by Robert Boyle in his will to defend religion from atheism, and consulted Newton on the "frame and origin of the universe". Newton replied at length, pointing out among other things that "... the motions which the Planets now have could not spring from any naturall cause alone but were impressed by an intelligent Agent." In other words, the current constitution of the solar system is proof for the existence of God, a conclusion that later prompted the philosopher R.G. Collingwood (1889–1943) to rebuke Newton for "...exalting the limitations of his own method into a proof of the existence of God."[14]

---

[11] Laplace, P.-S. (1798–1825). Translated into English by Somerville, M. (1831). Her translation was praised by Laplace.

[12] Bentley attended Newton's lectures as an undergraduate at Cambridge between 1676–80.

[13] Newton, I., (1692), p 331.

[14] Collingwood, R. G. (1960), p 109.

Collingwood's words could well have been uttered by Laplace because in 1786 he succeeded in proving that, within the limits of what was known about the solar system in his day, the irregularities in planetary orbits that had so troubled Newton were, given sufficient time, self-correcting and that consequently the present solar system is stable.[15] Hence Laplace's wry answer when asked by Napoleon why there was no mention of God in his Méchanique Céleste: "Sire, I have no need for that hypotheses."[16] Indeed, Laplace claimed that were it possible to know the position and velocity of every particle in nature at a particular instant, it would be possible in principle to predict all future events, though he acknowledged that such a task was beyond human powers. But he was merely giving voice to the unquenchable French faith in reason and logic that had inspired Descartes and was the inspiration of French science in Laplace's day. Bolstered by his successful account of planetary dynamics, Laplace and his circle looked forward to creating an equally rigorous mathematical account of optical phenomena based on the corpuscular theory of light.

Across the Channel there were no British mathematicians to equal those in France, but there was plenty of native scientific talent to rival the best that France had to offer. However, science in Britain was a far more laissez-faire business than it was in France, as it had been since the foundation in the seventeenth century of the Royal Society and the Académie Royale des Sciences.

As we saw in the last chapter, the Académie Royale was state-funded and its membership limited to a handful of savants selected for their expertise and who were, in effect, civil servants. There was no place in the Académie for the dilettante gentlemen that the Royal Society—which received no state support—was forced to admit to its ranks to pay the bills and ensure its survival. Although the Académie Royale was abolished in 1793, at the height of the French Revolution—it was considered to be an elitist institution and its incumbents hostile to the revolution—it was replaced in 1796 by l'Institut de France. The new organisation had a wider remit than the Académie and consisted of three sections known as Classes. The largest of these was concerned with the physical sciences and was known as the First Class.[17]

---

[15] Recent calculations indicate that the motion of the planets can't be accurately predicted more than 100 million years in advance.

[16] Laplace first came across Napoleon in September, 1785, when he was an examiner in mathematics at the École Militaire in Paris. He ranked the future emperor 42 out of a class of 58.

[17] The First Class of l'Institut was equivalent to the old Académie Royale des Science and was limited to 60 members. The Second Class was concerned with moral and political sciences and had 36 members and the Third Class with 48 members was devoted to literature and fine arts.

And, following the practice of the Académie Royale, its members were all professional scientists or mathematicians.

As a result, in the early years of the nineteenth century, France had a large body of highly trained specialists. In Britain, on the other hand, science was neither as professional nor as specialised as it was in France; tellingly, all members of the Royal Society were styled "Fellows", whatever their metier. Where the French were either physicists or chemists or mineralogists the British were content to call themselves natural philosophers and turn their hand to whatever field of research took their fancy. Thus, despite their understandable loyalty to Newton and respect for the achievements of French science and mathematics, British natural philosophers had a reputation for being eclectic and pragmatic, none more so than the brilliant polymath, Thomas Young (1773–1829).

Young was born in Milverton, a village in Somerset. His parents were strict Quakers, which may have had something to do with his somewhat stiff manner as an adult, long after he had ceased to be a practicing Quaker. He was an archetypal child prodigy, an awesome autodidact for whom no subject was off limits. He learned to read before he was two and by his fourth birthday had read the Bible all the way through, not once but twice—with, it has to be said, adult help. His formal education ended at thirteen, by which time he was proficient in several languages and had a well-developed interest in the sciences, particularly optics.

For the next five years he was engaged as a companion and tutor to Hudson Gurney (1775–1864), two years his junior, who was to become his closest friend. Like many prodigies, Young needed no teacher himself; indeed, he preferred things that way: "...whoever would arrive at excellence must be self-taught", he once informed his brother. He read widely in Latin, Greek, English, French and Italian, and worked through the major scientific works of the day including Newton's *Principia* and *Opticks*. Although his first love was languages, especially Oriental languages, he was equally at home in the sciences and mathematics. But when he came to choose a career he was persuaded by his mother's uncle, Dr Richard Brocklesby (1722–1797), to study medicine.[18]

Brocklesby died the year before Young graduated, but he left him a small fortune and a house in London, which allowed the newly minted Doctor Young to set up in private practice while affording him the leisure to pursue his scientific interests. Despite his formidable gifts, however, he struggled to

---

[18] Medicine was one of the few university courses involving science available to Englishmen at the time. The Scots were far better served in this respect.

succeed in his chosen profession. His patients complained about his perfunctory manner and lack of sympathy and as a result he never made his mark as a physician despite the many important technical contributions he made to medical science. A few years later, in 1804, he married Eliza Maxwell (1785–1859). It was, he said, a happy union. It also provided him with an extended family, which may have compensated for their childless marriage, for he became extremely attached to all three of Eliza's sisters. Indeed, he is buried in the Maxwell family vault in Farnborough, Kent.[19]

In 1801 Young was hired at the behest of Sir Benjamin Thomson, Count Rumford (1753–1814), the prime mover in the creation of the recently founded Royal Institution, as its Professor of Natural Philosophy. Thomson, or Rumford as he is more usually known, was an American who had sided with the British during the American War of Independence and had abandoned his wife and child when he moved to Britain at the end of the war. He later acquired a title for services as Minister of War to the Elector of Bavaria and had somehow persuaded George III to award him a knighthood—all of which contributed to his reputation as an incorrigible opportunist. But he was also known for his skill as an organiser and as an accomplished applied scientist. Moreover, during his time in Bavaria he had made major discoveries concerning the nature of heat, the most important of which was to recognise that heat is generated through friction when he noticed that cannon barrels manufactured in the Elector's Munich arsenal became extremely hot as they were bored. At the time it was widely held that heat is a fluid, given the name caloric by Antoine Lavoisier (1743–1794), the leading chemist of his day and the acknowledged founder of modern chemistry. According to this theory, as the drill bit rotated it squeezed caloric from the metal of the barrel, like water being wrung from a wet cloth. Rumford pointed out that as long as the process of boring was maintained, seemingly limitless quantities of heat could be produced and proposed that therefore heat is a form of motion brought about by friction rather than a substance locked up in matter, an idea that was not immediately accepted.

Rumford was better known as a prolific and ingenious inventor, particularly of domestic devices, and for his ideas for increasing the efficiency of kitchens, fireplaces and chimneys, all which made his name in England.[20] So, although he was an active Fellow of the Royal Society, as an applied scientist he saw the need for an institution devoted to teaching applied sciences that would complement the ivory tower that was the Royal Society. Young

---

[19] There is a tablet in his memory in Westminster Abbey. The inscription is by his great friend, Hudson Gurney (1775–1864).

[20] Thomas, J.M. (1999).

was to lecture on natural philosophy and Humphry Davy (1778–1829), a young Cornishman who had recently become famous for his discovery of the intoxicating effects of nitrous oxide, known popularly as 'laughing gas', was employed to lecture on chemistry. Davy was largely self-taught and had learned much his chemistry from Lavoisier's influential *Traité Élémentaire de Chimie*.[21] Rumford, who had recommended Young to the managers of the Royal Institution without reservation, had second thoughts about Davy when they first met: the young man struck him as a rough diamond, "uncouth in appearance and dress". He insisted that Davy give a private demonstration of his lecturing skills before endorsing him; Davy passed the test with flying colours. It soon became apparent that Rumford should have demanded the same of Young.

Young's lectures turned out to be the most comprehensive review of scientific knowledge of his day, but were largely beyond the typical Royal Institution audience of "silly women and dilettanti philosophers",[22] most of whom attended in expectation of being amused rather than instructed. Inevitably, audience numbers fell away as people discovered that Young was taking his brief in earnest: his lectures were little short of an advanced course in physics and astronomy, with only the difficult mathematics left out. And as if this wasn't enough to tax his audience, some lectures included accounts of his own theories and discoveries including the earliest use of the concept of energy in its modern sense of a measure of the capacity to do work,[23] the first reliable estimate of the size of a molecule,[24] and a new approach to the strength and elasticity of solid materials.[25]

But the scientific discovery for which he is principally remembered today formed the subject of a lecture on optics in which he dealt with the nature of light and colour in terms of waves rather than particles. The lecture included a detailed account of the experimental evidence for the wave theory of light and an explanation for supernumerary arcs based on what he called his "principle of interference".[26] No natural philosopher or mathematician of the time, British or French, would have fully understood or been sympathetic to any of these ideas, so we can safely assume that this lecture would have been well-nigh incomprehensible to his audience.

---

[21] Lavoisier, A. L. (1789).

[22] Peacock, G. (1885), p 118.

[23] Young, T. (1845), pp 59–60.

[24] Young, T. (1845), p 466–7.

[25] Young, T. (1845), p 106.

[26] Young, T. (1845), p 367.

Frustrated by what he saw as a betrayal of Rumford's ideals by the managers of the Royal Institution, who had been forced by the parlous state of its finances to make income-generating popular lectures a priority over scientific education, Young resigned his position in 1803. Davy stayed on for another ten years during which he made several important discoveries in chemistry and electricity, invented the miners' safety lamp and established the Royal Institution as Europe's leading centre of scientific research.[27] He was also one of the first chemists to exploit the potential of the electric battery, invented in 1800 by Alessandro Volta (1745–1827), as a scientific tool, having concluded that electricity is an essential property of matter and used it to isolate the elements potassium and sodium in 1807.

But Young faced a greater impediment to his scientific ambitions than the loss of his lectureship, one that dogged him throughout his life and which seems to have worked against the acceptance of his ideas, particularly in optics: the style of his delivery. His lectures lacked spontaneity and he was often laconic to the point of being Delphic, his vocabulary unfamiliar or obscure. He freely acknowledged that Davy was a more successful lecturer but was himself unable to pitch his lectures to suit the typical Royal Institution audience. At Cambridge, where he had completed his medical studies, his fellow students had regarded him with an uneasy mixture of respect and ridicule and nicknamed him "Phenomenon Young". A Cambridge contemporary remarked that Young was "worse calculated than any man I ever knew for the communication of knowledge…for he presumed…on the knowledge and not on the ignorance of the hearers."[28] In all likelihood Young suffered from a mild form of Asperger's syndrome, an condition that some psychologists in our day appear to regard as the sine qua non of scientific genius, for it was said that he "…never could either make a joke or understand one", something that might, of course, be partly attributable to his straight-laced upbringing.[29] Young himself admitted that he found it difficult to read between the lines when confronted with a "…recital which the narrator had no intention whatever to impress on his audience as a matter of fact."[30]

Humphry Davy, on the other hand, as well as being an accomplished scientist with a string of important discoveries to his name, was also a natural showman. His lectures drew large, enthusiastic crowds and became a fixture

---

[27] More elements have been discovered at the Royal Institution than at any other research laboratory bar one, Laurence Berkley National Laboratory, where 16 elements have been found. Davy himself discovered 7 elements and had a hand in identifying several others.

[28] Peacock, G. (1855), p 118.

[29] Peacock, G. (1855), p 117.

[30] Hilts, V.L. (1978), p 254.

of the London Season with high society. The lecture hall of the Royal Institution could accommodate up to 1000 people and he had no difficulty in filling it, which led to Albemarle Street, on which the Royal Institution is situated, becoming the first street in London to be made one-way to deal with the crush of traffic whenever Davy was lecturing. Such was his celebrity that when he fell ill in 1807, the managers of the Royal Institution were forced to post hourly updates of the state of his health outside the entrance to the building to satisfy public concern. In fact, Davy's lectures were the lifeblood of the Royal Institution in its early years. Without the income they generated it is doubtful that the Institution would have survived to house its most illustrious incumbent and Davy's successor, Michael Faraday. However, the principal attraction for the large number of young women who attended these lectures was not chemistry but Davy himself: his charm, vivacity and good looks were irresistible to them. One of his legions of female admirers declared that that "…those eyes were made for something besides poring over crucibles".[31]

No sooner had Young begun training as a doctor in London in 1792 than he embarked on a study of the human eye, concentrating on the nature of mechanism by which it alters its focus as one's gaze shifts from a distant object to one near at hand or vice versa, a process known as *accommodation*. The following year he discovered that this is accomplished by changes in the curvature of the lens rather than that of the cornea, which was the prevailing opinion at the time. On the strength of his discovery, he was elected a Fellow of the Royal Society in 1794, having just turned twenty-one. Several years later, in 1801, he realised that the perception of colour can't depend on each receptor in the retina being sensitive to every part of the spectrum. Instead, he suggested, only three types of receptors are necessary: one sensitive to red, another to yellow and a third to blue; he later changed these to red, green and blue. But he left it to others to develop these novel ideas on colour perception.[32]

Here is yet another reason why Young's discoveries were often ignored or overlooked by his contemporaries: he himself summed it up admirably "…acute suggestion was … always more in the line of my ambition than experimental illustration."[33] He seldom pursued his researches to their logical conclusion and, despite being an able mathematician himself, had no appetite for the rigorous mathematical physics of the French. He confessed to a friend that "…were I to apply deeply [to mathematics], I would become a disciple

---

[31] Williams, L. P. (1965), p 19.
[32] Young, T. (1802 b), pp 20–1.
[33] Peacock, G. (1855), p 397.

of the French and German school; but the field is too wide and too barren for me."[34] Yet, as he knew only too well, the continental school of mathematics was what enabled French scientists to outclass their British counterparts in the theoretical aspects of science at the turn of the nineteenth century and which, as we shall see, made it possible for Augustin Fresnel (1788–1827), destined to be Young's unwitting French rival in optics, to leapfrog him and arrive at an exhaustive mathematical account of light as a wave that was eventually adopted by all but the most diehard Newtonians.

In view of Fresnel's success, it is an irony that Young was probably the first Englishman to acknowledge "…how much the foreign mathematicians for these forty years have surpassed the English in the higher branches of sciences. Euler, Bernoulli, and D'Alembert have given solutions of problems which have scarcely occurred to us in this country."[35] But his pioneering scientific work, which should have acted as call to arms for his countrymen, fell upon deaf ears and British science did not fully awaken from its one-hundred-year Newtonian slumber for another generation.[36]

One of Young's acute suggestions was also his greatest contribution to physics: the principle of interference. The idea was sparked off by private research into sound while he was at Cambridge University, where he had to spend a couple of years in to comply with the requirements of the London-based College of Physicians. Although he had already qualified as a doctor from Göttingen University, where the subject of his doctoral thesis was the human voice, in order to gain a licence to practice medicine in London he had to spend two consecutive years at the same institution—he had spent only a year at Edinburgh. and another at Göttingen. Thus, during his time at Cambridge he was free to do as he wished and devoted himself to investigating the physics of sound. As a result of these investigations he became convinced that there are close similarities between certain acoustic and optical phenomena.[37] The most telling of these is the phenomenon of beats, the result of combining two pure tones of similar loudness that differ slightly in pitch. The net effect is a tone that regularly rises and falls in loudness (i.e. pulsates) because the vibrations from the two sources are alternately in step (loud sound) and out of step (quiet sound).

---

[34] Peacock, G. (1855), p 127.

[35] Young, T (1798). In: Dalzel, A. (1862), p 161.

[36] The young Turks of the revival were John Herschel (1792–1871), Charles Babbage (1791–1871) and George Peacock (1791–1858), who, when they were undergraduates at Cambridge, founded the Analytical Society in 1812 with the aim of promoting the use of Continental mathematical methods in place of Newtonian fluxions.

[37] Steffens, H.J. (1977), pp 109–10.

Musicians had been using the phenomenon since the early sixteenth century to tune organs by ear, because tones that have the same frequency do not beat. But no one before Young had understood what causes beating. He realised that the rise and fall in loudness was the result of two tones reinforcing one another when their vibrations coincide and cancelling one another when they are out of step. Young reasoned that the same thing should occur in the case of light considered as a wave: where light waves are in step, the result is brightness; where they are out of step, it is darkness.

Young claimed that he discovered the principle of interference in 1801, not through experiment but "...by reflecting on the beautiful experiments of Newton."[38] Assuming that the series of bright and dark bands of Newton's Rings are due to two reflections that are either in step or out of step with one another, he used Newton's measurements of the thickness of the film of air corresponding to each colour in the spectrum to calculate their wavelength and frequency, adding that it was only later that he came across Hooke's account of colours in thin films in *Micrographia*. Had he known of Hooke's work beforehand, he said, he might have been led to the principle of interference sooner: "It was not till I had satisfied myself respecting all these phenomena, that I found in Hooke's *Micrographia*, a passage which might have led me earlier to a similar conclusion."[39] Young's calculations of the wavelengths of coloured light are remarkably close to those accepted today and are a testament as much to the astonishing accuracy of Newton's measurements, made 130 years earlier, as they are to Young's scientific intuition.

But Young's innovative reworking of Newton's data, which attracted not a scintilla of interest at the time, hardly amounted to a proof of the principle of interference or, by extension, of the hypothesis that light is a wave rather than a particle. What was needed were experimental results that could be explained only in terms of waves, something that took him a further two years to achieve. The most convincing of these experiments took place in a darkened room and involved splitting a narrow beam of light from a candle by passing it through an opaque screen in which two small pinholes had been made very close together. He found that where the light from both pinholes overlaps there is a pattern of alternating bright and dark patches (Fig. 8.1).

Young explained these observations as follows: bright patches occur where light waves from one pinhole are in step with those from the other pinhole and dark ones where they are out of step. Moreover, the pattern of bright and dark patches repeats because bright patches occur where the distance

---

[38] Young, T. (1855), p 202.
[39] Young, T., (1802 a), p 39.

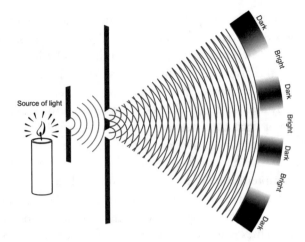

Fig. 8.1 Young's interference patterns (one of the most important and influential experiments ever performed). Where light waves are completely out of step the cancel out and the result is darkness

from one pinhole is either equal to that from the other or differs from it by a whole number of "undulations", his word for wavelengths. Where the difference is a half wavelength the waves are completely out of step and the result is darkness. This is a startlingly counterintuitive result, for it appears that in the right circumstances combining light from two sources it is possible to produce darkness.

> The middle of the two portions is always light, and the bright stripes on each side are at such distances, that the light coming to them from one of the apertures must have passed through a longer space than that which comes from the other, by an interval which is equal to the breadth of one, two, three or more, of the supposed undulations, while the intervening dark spaces correspond to a difference of half a supposed undulation, of one and a half, of two and a half, or more.[40]

In fact, it is only with monochromatic light that the dark patches really are dark. Candlelight, like sunlight, consists of all the spectral colours, though in candlelight the blue end is far less bright than it is in sunlight. On closer inspection Young noticed that the bands were, in fact, "a beautiful diversity of tints, passing by degrees into each another."[41] You will have seen Young's

[40] Young, T. (1845), p 365.
[41] Young, T. (1845), p 365.

Fig. 8.2 A close up photograph of several supernumerary arcs next to the blue arc of a rainbow[42]

"diversity of tints" whenever sunlight passes through your eyelashes when your eyes are all but closed.

Young went on to show how the principle of interference can be used to explain the colours seen in thin transparent films such as soap bubbles, in Newton's Rings and in finely scratched surfaces. But perhaps the most unexpected application of the principle was his explanation of supernumerary arcs, for it is not immediately obvious how the two closely spaced beams necessary for interference come about within a drop (Fig. 8.2).

Young had realised that according to Descartes' geometrical explanation of the rainbow, rays emerging from a drop either side of the rainbow bow ray do so as parallel pairs. Moreover, the distance a given ray travels through a drop differs slightly from that of its immediate neighbours so that each pair of rays when considered as waves will either be in step or out of step as they emerge from the drop, the very condition necessary for interference to occur. If the difference in the distance travelled through a drop is equal to a whole number of wavelengths, the rays reinforce one another and give rise to a bright band. If the difference is equal to half a wavelength, the result is darkness. However, sunlight is composed of a broad range of wavelengths, so supernumerary arcs do not consist of alternating bands of brightness and darkness. Just as happens with Newton's Rings, the bright and dark bands of one colour overlap those of another giving rise to bands of intermediate colours such as pink (red + violet) or green (yellow + blue). Hence the series of narrow, concentric pink and green arcs sometimes seen just beyond the blue arc of the primary bow (Fig. 8.3).

---

[42] Photo by Mika-Pekka Markkanen, Wikimedia Commons. https://tinyurl.com/ym85s993.

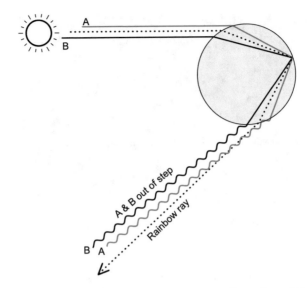

Fig. 8.3 Young's explanation for supernumerary arcs. Rays A and B travel slightly different distances through the drop and emerge from it either in step or out of step, which results in either constructive or destructive interference

Young also claimed that "…unless almost all the drops of the shower happen to be of the same magnitude, the effects … must be confounded and destroyed."[43] In other words, supernumerary arcs will not be seen if the size of drops is not uniform. Moreover, only the smallest drops will do: "The magnitude of the drops of rain, required for producing such … rainbows … is between the 60th and the 100th of an inch".[44] But, surprisingly, he offered no detailed explanation for any of this. Nor did he bother to explain how he had arrived at the size of these drops, all of which left his account of supernumerary arcs incomplete.[45]

The Royal Institution lectures were not the only occasions on which Young made known his ideas on light. He covered the same ground when he delivered the Bakerian Lecture at the Royal Society on three successive years: 1801, 1802 and 1803. Unfortunately, as we have already noted, in the early 1800's the British scientific establishment lacked anyone who could appreciate the significance of his ideas on optics, let alone pass informed comment. In any case, the Royal Society was a shadow of its former self, having become something of a gentleman's club,[46] its membership still open to gentlemen

[43] Young, T. (1845), p 369.

[44] Young, T., (1845), p 369.

[45] For Young's explanation for supernumerary bows see Young, T. (1804), pp 8–9.

[46] Babbage, C. (1830), pp 50–2.

untrained in any science and presided over by Sir Joseph Banks (1743–1820), who considered the use of mathematics in any scientific paper to be undesirable.[47] Moreover, at that time it was deemed bad form to comment publicly on papers read at meetings at the Royal Society because such discussions "led to the loss of personal dignity."[48] Given these circumstances, it is little wonder that Young's ideas on light made hardly any impact on the scientific establishment of the day.

In the meantime, Young had not endeared himself to some of his peers through his habit of speaking his mind without regard to consequences. His most implacable adversary was Henry Brougham (1778–1868), whose amateur efforts to demonstrate a talent for mathematics had been the object of one of Young's unsparing appraisals. Brougham, a prodigiously energetic and opinionated lawyer who aspired to be known as a polymath, and who was an uncritical devotee of Newtonian corpuscular theory into the bargain, now saw an opportunity to strike back. He wrote a series of anonymous articles for the *Edinburgh Review*, one of the most widely read and influential journals of the time, attacking Young's lectures at the Royal Institution and the Royal Society. He criticised their content and heaped abuse on their author.

> We now dismiss, for the present, the feeble lucubrations of this author, in which we have searched without success for some traces of learning, acuteness, and ingenuity, that might compensate his evident deficiency in the powers of solid thinking, calm and patient investigation, and successful development of the laws of Nature, by steady and modest observation of her operations.[49]

The year before he had twisted the knife by suggesting that Young's ideas were no more than "fashionable theories for the ladies who attend the Royal Institution."[50] If only Brougham had consulted those ladies he might have been taken aback to discover how few were likely to have been present during Young's discourse on light.

Incredibly, given that none of Brougham's articles contained a single substantive refutation of any aspect of Young's wave theory or of his experimental evidence for the interference of light, these intemperate attacks damaged Young's scientific reputation and delayed the publication of his lectures. Young worried that they threatened his medical reputation. His attempt to defend himself fell on deaf ears, as a consequence of which, in

---

[47] Members of the R.S. were always referred to as "Fellows", never as "Natural Philosophers"; and certainly not as chemists, biologists or physicists.

[48] Boas Hall, M. (1984), p 69.

[49] Edinburgh Review, vol. 5, Oct 1804, p 103.

[50] Edinburgh Review, vol. 1, Jan 1803, p 452.

echoes of Newton, Young proclaimed that he would abandon all further scientific research and henceforth devote himself to exclusively to languages and medicine.[51] And given that Young was then its only advocate, the wave theory of light was destined to languish unappreciated and ignored for several years to come.

France proved equally stony ground for Young's ideas. As we have noted, the French scientific establishment of the day had embraced Newtonian physics with the single-minded devotion to abstract principles that at times has characterised the French intellect. In their hands, Newton's dynamics and gravitational theory had been refined and improved to the point where it provided a mathematically rigorous and seemingly comprehensive account of planetary motion. And despite the lack of an equivalent success in Newtonian optics, they were by default almost to a man, fervent advocates of the corpuscular theory of light and vehement opponents of any theory of light that involved vibration. Not that they were unaware of Young's wave theory. Although France and Britain were at war with one another for most of the first decade and a half of the nineteenth century, even at the height of hostilities copies of the *Philosophical Transactions* seldom failed to reach Paris.

If Young's "lucubrations" on the subject of light made no more impression in France than they did in Britain, the very issue of the *Philosophical Transactions* in which Young published his principle of interference also carried details of an investigation that, in the opinion of Laplace, threatened to breathe new life into the detested hypothesis of luminiferous waves.[52] This was the experimental confirmation by one of Britain's most skilful chemists, William Hyde Wollaston (1766–1828), of Huygens' geometrical explanation for double refraction in Iceland spar.[53] As a result, spurred on by Laplace, in January, 1808, the First Class of the l'Institut de France proposed a prize for an essay on nature of double refraction, in the hope that it would produce an account of the phenomenon based firmly on corpuscular principles.

The prize was won by Étienne Louis Malus (1775–1812), a military engineer and ardent Laplacian who had taken part in Napoleon's abortive 1798 expedition to Egypt. A few months after arriving in Egypt, Malus was dispatched to Syria where he was involved in the siege and capture of Jaffa, the savagery of which marked a low water mark in the history of French arms. Soon after, most of the occupying force fell prey to plague and died. Malus also fell ill and survived only by the skin of his teeth, as a result of which his health never fully recovered and which contributed to his untimely death

---

[51] Young, T. (1855), p 215.
[52] Young, T. (1802 b), pp 387–97.
[53] Wollaston, W.H. (1802).

from tuberculosis. On his return to France he resumed his military career. He was famously reserved and laconic, having, according to a contemporary, "a taste for silence"; when he was an examiner for the École Polytechnique, the only indication he would give his students of mistakes in their work was to tap the offending passage with his finger.[54]

Malus had a long-standing interest in optics dating back to his time in Egypt, so when the prize essay was announced he immediately set to work and within a short time made a great discovery. Late one afternoon, in 1808, while looking idly through a crystal of Iceland spar at a reflection of the setting sun in one of the glass windows of the Luxemburg Palace he was surprised to find that, depending on the orientation of the crystal, he saw only one bright image of the sun where he expected to see two.[55] His immediate thought was that the sun's reflection was affected in some unknown way by the intervening atmosphere. But further experiments that night convinced him that this was not the case: the reflection of a candle flame in a dish of water viewed through the crystal produced identical results to that of the sun's reflection. He drew the only possible conclusion: light is affected by reflection in the same way as it is by refraction through Iceland spar. Being a committed Newtonian, he suggested that Newton's "sides" could be explained by assuming that corpuscles of light have poles like a magnet. When the poles of all the corpuscles in a beam of light are lined up in the same direction, said Malus, the beam is polarised.[56] His experiments not only proved that light is polarised when it is reflected and confirmed Huygens' account of the extraordinary ray, he also found that the degree of polarisation depends on the angle of reflection as well as on the substance of the reflecting surface, being greatest in the case of water at 53° and 56° for glass. At the same time Malus also established that the portion of light that is refracted when it encounters a transparent body such as glass or water is also polarised, though in the opposite sense to that of the light that is reflected. He communicated his findings to the Academy later that year and published them the following year (Fig. 8.4).

Malus' discovery was seized upon by François Arago (1786–1853), a young astronomer who was later to play a pivotal role in furthering the cause of the wave theory of light. Arago was then the youngest and newest member of l'Institut and saw polarisation as a field in which he could make

---

[54] Arago, F. (1857), p 394.

[55] In fact, Malus found that neither of the reflected images of the sun vanished completely as he rotated the crystal because the sun's light was not, in this case, given the angle at which it was reflected by the glass, completely polarised by reflection.

[56] Malus coined the term *polarisation*.

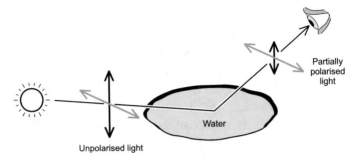

Fig. 8.4 Polarisation by reflection. Light polarised in a plane parallel to a reflecting surface is preferentially reflected and light polarised perpendicularly to a surface is absorbed. Polarisation by reflection is greatest when the angle of incidence is equal to Brewster's angle

a name for himself. Holding a sheet of mica up to the sky and examining it through Iceland spar in order to investigate the colours that are brought about by polarisation in some circumstances, Arago inadvertently discovered that skylight is strongly polarised in a direction perpendicular to the sun's rays. He already knew that these colours could only be seen when the light passing through the sheet of mica is polarised, hence, he realised, skylight must be polarised.[57] These and other optical discoveries led to a dispute over priority with a fellow Academician, the physicist Jean-Baptiste Biot (1774–1862). Biot was one of Laplace's men and although l'Institut eventually found in favour of Arago, he never forgave Biot, which may have played a part in his later support for the wave theory. Biot also made optical discoveries: early in 1811, while studying polarisation in reflections from the surface of different substances he discovered that the arc of the rainbow is strongly polarised.[58] The following year, David Brewster (1781–1868), a Scottish physicist who went on to become a leading authority on optics and polarisation, unaware of Biot's discovery, looked at a rainbow through Iceland spar.

> Upon examining with a prism of Iceland crystal the light of a very brilliant rainbow, I was surprised to find, that one of the images of the coloured arch alternately vanished and re-appeared in every quadrant of the circular motion of the prism. The light, therefore, which forms the bow has been almost wholly polarised…[59]

[57] Arago, F. (1858), p 394.
[58] Biot, J.-P. (1811), pp 282–3.
[59] Brewster, D. (1813), p 350.

That the light of a rainbow is polarised is to be expected because it involves reflection. And, as Malus had discovered, depending on the angle of reflection, reflected light is usually polarised to a greater or lesser extent. Moreover, by a happy coincidence, the average angle of reflection of the rainbow rays responsible for the primary arc of all spectral colours within a drop is approximately 40°, which is very close to the angle for which polarisation by reflection is at a maximum for white light at the interface between water and air—i.e. within water—which is approximately 37°.[60] As a consequence, the light of the primary rainbow is almost completely polarised. Indeed, we now know that the primary rainbow only just misses out being the most highly polarised source of light in nature: the degree of polarisation of the light from the primary bow is 95% and that of the secondary bow is 90%.[61] By contrast, the polarisation of skylight is never more than 70%.

Unlike colour and brightness, however, the polarisation of light is all but imperceptible to the human eye, which is why this most fundamental optical property went unnoticed until Huygens carried out his investigation of double refraction.[62] Even then, as we saw in the last chapter, it was considered a minor phenomenon, an unusual property confined to Iceland spar, not an intrinsic quality of light like brightness or colour. Newton had guessed correctly that double refraction is due to an inherent property of light but no one followed up this suggestion. So it was only when Malus, Arago, Biot and others devised better methods of detecting polarised light that it became apparent that, except for direct sunlight, almost all light in nature is polarised to some degree.[63]

Fortunately, since those early days, methods for detecting polarised light have become much less cumbersome. Moreover, modern polarising filters are both cheap and readily available: they are used in polarising sunglasses and as filters for camera lenses because, when suitably aligned, they reduce unwanted reflections. This type of polarising filter reveals the presence of polarised light through a change in the brightness of a surface as the filter through which it is viewed is rotated. And because a rainbow is so highly polarised, portions of it can be made to vanish when looked at through a polarising filter that is suitably orientated. The reason why the entire rainbow does not vanish when

---

[60] The angle of reflection that produces maximum polarisation for any two media is given by Brewster's law.

[61] Only a parhelion is more highly polarised, though its polarisation is not as obvious as that of a rainbow.

[62] But see remarks on Haidinger's Brush in the appendix.

[63] This is because most of the light reaching our eyes is due to reflection from the surfaces of objects around us. The only source of unpolarised light in nature is direct sunlight and starlight. See: Naylor (2002), p 25–8.

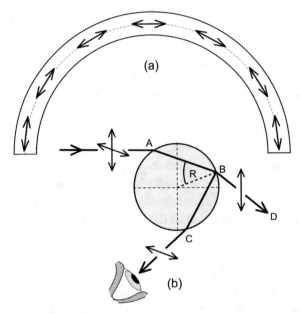

Fig. 8.5 Tangential polarisation of light from a rainbow. **a** Light from the primary bow is polarised in a plane that is perpendicular to the bow's radius as indicated by the double-headed arrows within the arc. This fact is only noticeable when the bow is viewed through a polarising filter. **b** Light entering the drop at A is unpolarised and is reflected within the drop at B very close to the Brewster angle R, so emerges at C almost entirely polarised in a direction that is perpendicular to that which passes through the drop into the air beyond in the direction D

viewed through a polarising filter is that the bow is polarised in a direction that follows the curve of the arc.[64] Light from the apex of the bow is thus polarised in a direction perpendicular to that from the foot of the bow. Thus, if a filter is aligned so as to absorb polarised light from the apex of the bow, polarised light from the foot will still be visible (Fig. 8.5).

The discovery that polarisation is an inherent property of light was a serious blow to early versions of the wave theory. In 1811 a disconcerted Young wrote to Malus "Your experiments demonstrate the insufficiency of a theory [of interference] which I had adopted, but they do not prove its falsity."[65] Malus was delighted: the leading advocate of the wave theory was expressing doubts! Young's uncertainty stemmed from the fact that if, as he believed, light waves are similar to sound waves, it is difficult to account for polarisation. Sound waves are longitudinal: the vibration of the medium

---

[64] Physicists call this "tangential polarisation".

[65] Arago, F. (1857), p 390.

through which they travel is back and forth in the same plane as the direction of propagation of the wave; and as Young knew, such waves cannot be polarised. But Malus' triumph was not to last for it merely stimulated Young to look for a way around the problem. In 1817 he suggested tentatively that light might be a transverse wave.[66] In other words, the medium through which the wave travels vibrates at right angles to the direction of propagation. By then Malus had been dead five years and the corpuscular theory had been put on final notice by a brilliant young French engineer, Augustin Fresnel.

Augustin Fresnel was the son of Jacques Fresnel, an architect who at the time of his son's birth was overseeing repairs to buildings on the estate of Marshall Victor-François de Broglie—an irony relished by all physicists who know of it, the significance of which will be explained in the next chapter. His mother was Augustine Mérimée (1755–1833), sister of the painter Léonor Mérimée (1757–1836), who later became France's leading authority on the chemistry and technology of paint.[67] A year after Augustin's birth in 1788, the Fresnels moved back to Mathieu, their home village in Normandy.

In contrast to the disconcertingly precocious Thomas Young, Augustin Fresnel was a slow starter. It seems that he didn't learn to read fluently until he was eight years old and was considered something of a dullard by his teachers. But to his peers he was an incomparable "l'homme de genie", the man of genius. Like all great experimentalists, his dexterity and practical ingenuity manifested themselves while he was still in short trousers; in his case it took the form of ballistics. Country children in those days had to make do with whatever was at hand by way of entertainment. Popguns, irresistible to young boys, were made from hollowed-out branches of elder wood and the necessary pellets from the pith. The pellets were fired by rapidly ramming a tightly fitting twig into the wooden barrel and could inflict a nasty weal on a victim's exposed flesh. By experimenting with different combinations of length and bore, Augustin created weapons of such efficiency that he and his classmates were victorious in every encounter with rival groups. Things quickly got out of hand and parental protests eventually put a stop to Augustine's missile experiments.[68]

The remainder of his childhood and adolescence passed without further incident, possibly as a result of a frail constitution brought about by tuberculosis. In 1804 he was admitted to the École Polytechnique, having impressed

---

[66] Young, T. (1855), p 383.

[67] Léonor was father to Prosper Merimée (1803–70), the writer and cultural historian, the author of the short story on which Bizet based his opera 'Carmen'.

[68] Arago, F. (1857), p 402.

his examiners with his knowledge of geometry, a subject in which he excelled. The Polytechnique, which was founded in 1794 during the French Revolution and militarised by Napoleon the year that Fresnel was admitted, provided budding engineers, civil and military, with a rigorous technical and mathematical grounding before they transferred to specialist schools. But until it was reformed by Napoleon, attendance of lectures was patchy and Parisians complained constantly about the rowdiness and indiscipline of its students. However, Napoleon's primary interest in the Polytechnique was as a source of technically proficient cannon fodder, not as hothouse for scientists and engineers. Under the new regime, students were required to reside in barracks, wear a military uniform, complete with a sword and a jaunty bicorn cap, and to march in formation to and from class preceded by a drummer and under the command of an officer cadet.

Fresnel eventually graduated as a civil engineer from the École des Ponts et Chaussées in 1811 and was immediately posted to the Vendée, a Department on the French Atlantic coast, to supervise the building of roads, a role for which he was singularly unfitted.[69] "Je ne trouve rien de si pénible que d'avoir à mener des hommes"[70] he complained—"There is nothing I loathe more than having to lead men." His frail constitution made travelling around the countryside on horseback on site visits an ordeal. Above all, he was bored; road-building may have been intellectually undemanding, but it left him with little time or energy to do anything else.

Relief came in the form of a promotion and a fresh assignment. He was sent to Nyons in the Rhône valley, to a post that gave him sufficient leisure to pursue his scientific interests. Despite being cut off from the latest scientific developments, Fresnel was determined to make his name as a scientist. Fortunately, he had a lifeline to the wider world through uncle Léonor.

Fresnel's early scientific interests were chemical rather than physical, something that may have been influenced by a combination of Léonor's chemical expertise and the dire state of physics teaching at the Polytechnique. Its sole professor of physics, Jean-Henri Hassenfratz (1755–1827), was such a poor teacher that, before Napoleon's reforms were implemented, the few students who bothered to attend his lectures would regularly mock and heckle him.[71] Whatever the reason, Augustin and Léonor corresponded on topics such as recipes for inks and glues and discussed at length Augustin's ideas for a new

---

[69] The Vendée was the region that offered the greatest opposition to the French republican government in 1793. The brutal suppression of the revolt in which tens of thousands of unarmed civilians were systematically killed by the government is considered to be an early example of genocide. See: Burleigh, M. (2005). pp 97–100.

[70] Fresnel, A. (1866–70), n 2, p xxviii.

[71] Arago, F. (1857), p 7–8.

method for the manufacture of soda, which to the young man's dismay turned out to be already in use. Then, unexpectedly, his interest switched to optics, the field in which he was to make his name.

The first hint of his new enthusiasm was an admission of ignorance. In May, 1814, he wrote to Léonor: "I have seen in the *Moniteur* that, a few months ago, M. Biot has read at l'Institut a very interesting paper on the polarisation of light. I have tried hard, but cannot think what the word means."[72] This from the man who within three years was to revolutionise optics and wrest it once and for all from Newton's dead grasp and all but stop the Laplacian corpuscular juggernaut in its tracks, something that Young signally failed to do. He asked for books on optics to be sent to him, but the request went unanswered until early the following year. How Fresnel must have fretted as he waited and dreamed of devoting himself to science rather than road building.

The books, when they arrived, appear to have been of limited help. Far more useful was the enforced idleness of house arrest. Fresnel, though never a monarchist, was deeply anti-Bonapartist, and when Napoleon escaped from exile on Elba in spring 1815 and was making his way across France, Fresnel took down from above his mantelpiece the ceremonial sword he had worn as a cadet at the Polytechnique, dusted it off and hastened on horseback 400 kms west to Toulouse to join the forces loyal to the Bourbon king who had been placed on the French throne following Napoleon's exile to Elba in 1814. The exertion proved too much and he had to take to his bed. Forced to return to Nyons, he was cashiered and ordered to report to the police at regular inter-vals. A few months later he was granted permission to stay with his mother at the family home in Mathieu.

It was during his exile in Mathieu that Fresnel began to lay the mathe-matical foundations of the modern wave theory of light. Although he had no access to most of the scientific literature on the subject, Uncle Léonor had introduced him to Arago as he passed through Paris on his way to exile in Mathieu. Arago was by then a leading member of the Academy and some-thing of a French Halley: not only a versatile and able scientist, but affable, generous and adventurous.[73] Moreover, he was one of the few members of the Academy with an open mind on the question of the nature of light, being sceptical about Newton's ideas on the subject. In answer to Fresnel's inquiries, he suggested that he read Newton, Grimaldi and Young among others. Unfor-tunately, Fresnel knew not a word of English or Latin and, in any case, these works were available only in Paris, a city he was now forbidden to visit due

---

[72] Fresnel, A. (1866–70), n 1, p xxix.
[73] After the fall of Napoleon in 1815 the *First Class of l'Institut* was renamed *Académie des Science*.

to his act of insubordination. Thus, in almost complete ignorance of much of the current thinking on the nature of light, he was forced to strike out on his own, devising and performing experiments of extraordinary ingenuity and precision employing the crudest apparatus imaginable, some of which he had made by the village locksmith. To give some idea of just how primitive were the conditions under which he laboured, he solved the need for a tiny but powerful lens with which to focus the sun's rays by placing a drop of honey from his mother's bee hive in a small hole drilled in a thin metal plate. But in his skilled hands such improvised apparatus was enough to enable him to rediscover the principle of interference and convince him that light behaved like a wave rather than a particle.

Arago, who had kept abreast Fresnel's work since their meeting in Paris, now realised that such remarkable talent could not be allowed to squander itself repairing potholes in country roads. He must be allowed access to the libraries and laboratories of Paris. Arago pulled strings and had Fresnel released temporarily from duty and ordered to Paris in March 1816. At long last Fresnel was able to consult Young's papers on optics, and stimulated by contact with leading scientists and the opportunity to perform his experiments with professionally made apparatus, made huge strides in his research.

But the French state was not to be denied: after a few blissful months of uninterrupted research he was recalled to duty and put in charge of a workhouse, a task he found even more uncongenial than overseeing gangs of navies. Arago once again had to pull strings to get Fresnel back to Paris, if only for a few more months. All the while Fresnel's health worsened, though at the time no one could have known that within six years he would be too unwell to continue his scientific work. But by now, even the Ministry of Works had realised that Fresnel was no run-of-the-mill engineer and began to turn to him for advice on optical instruments. As a result Fresnel had a parallel career as Secretary to the Commission for Lighthouses, which led him to invent a type of flat, corrugated lens known as an echelon lens that is still used to create the narrow beam of light with which a lighthouse warns shipping and has since been adapted for use in overhead projectors and flat magnifying glasses.

Fresnel's earliest optical experiments were concerned with diffraction, which led him to the principle of interference. As we have seen, at the time Fresnel was unaware that this principle had been discovered and enunciated by Young more than a decade earlier, though he was more than happy to acknowledge Young's priority when he found this out. He wrote to Young

When one believes one has made a discovery one cannot learn without regret that one has been anticipated … But if anything could console me for not having the advantage of priority, it would be having been brought into contact with a scholar who has enriched physics with so many important discoveries and has contributed not a little to increase my confidence in the theory which I had adopted.[74]

But Fresnel had investigated the phenomenon far more thoroughly than had Young and, more importantly, drawing on the mathematics of waves and vibrations learned at the Polytechnique, he had developed a comprehensive mathematical theory of diffraction. To eliminate the possibility of Newtonian inflection (the idea that diffraction is due to attraction between matter and light), Fresnel devised an experiment in which the reflections of a single source of light by two suitably positioned mirrors are merged to create the tell-tale bright and dark interference bands. To explain these results mathematically, Fresnel combined the principle of interference with Huygens' principle of wavefronts. The resulting mathematical formulas enabled him to calculate the intensity of light at any point beyond an object, something that made it possible for him to account for the rectilinear propagation of light by showing that in most circumstances light within a shadow interferes destructively, i.e. leads to darkness. So much for Newton's central objection to the wave theory of light: that the absence of light beyond an edge is proof that it cannot be a wave.

These discoveries formed the substance of the essay that he submitted in 1819 to the Académie de Sciences in response to the prize the Academicians proposed to award for a mathematical and experimental explanation of diffraction based on the corpuscular theory.[75] As with the prize essay on double refraction that had been won by Malus, the Academicians hoped that this would produce a winning entry that would advance the cause of the corpuscular theory at the expense of the wave theory. Of the five members of the judging committee, three were dyed-in-the-wool Newtonians.[76] To ensure fairness, however, entries for the Academy's prize were always anonymous, the authors being identified only by a number. The 1819 competition attracted only two entries. The essay by the entrant identified as number one was considered so poor that the committee refused to consider it further. But the combination of experiment and mathematics of the other essay so

[74] Young, T. (1855), p 378.

[75] Fresnel, A. (1818), pp 339–475.

[76] The members of the panel were François Arago, Louis Joseph Gay-Lussac, Jean-Baptiste Biot, Pierre-Simon Laplace and Siméon Denis Poisson. Biot, Laplace and Poisson were in favour of the corpuscular theory.

impressed the judges that they unanimously awarded the prize to entrant number two, who was then revealed to be Fresnel.

One of the judges, Siméon Poisson (1781–1840), a supporter of the corpuscular theory, drew attention to what he considered a flaw in Fresnel's account: according to Fresnel's equations there should a bright spot at the centre of the shadow cast by a small disc. Experiments by Arago duly confirmed that this was indeed the case. But Poisson and his fellow corpuscularian judges were unmoved: in explaining their reasons for awarding the prize to Fresnel they completely ignored his explanations of diffraction in terms of waves and praised him only for his experiments and his mathematics.

Leading Newtonians like Biot and Poisson felt they could safely ignore the wave theory because they were confident that Fresnel had not provided sufficient evidence to call the corpuscular theory into question. In any case, the corpuscular theory appeared to offer a better explanation for polarisation than the wave theory. From the start, both Young and Fresnel had assumed that light is a longitudinal vibration, like sound. But as everyone acknowledged, a longitudinal wave cannot be polarised because it vibrates in the same plane in which it travels, i.e. back and forth. This is why Young had experienced "a descent from conviction to hesitancy" when he realised that Malus' discovery implied that polarisation is an inherent property of light. Fresnel, however, was undaunted. With Arago's help, he repeated his interference experiments with polarised light and discovered that interference occurs only when the two intersecting beams are polarised in the same sense. This was very strong evidence that light is a transverse vibration. Young, as we saw, had already reached the same conclusion the previous year on purely theoretical grounds.

But if the evidence that light waves are transverse provided an answer to the problem posed by polarisation, it introduced another. Everyone at the time assumed that all types of wave require a medium in which the necessary vibrations can occur. If light is a longitudinal wave, then the medium in which it travels, the æther, whatever its composition, should behave like a gas, the medium necessary for sound waves, i.e., it must be compressible. Unlike longitudinal waves, however, transverse waves require a rigid medium, i.e. a solid. Hence Young's agonised cry that the evidence that light is a transverse wave was "…perfectly appalling in its consequences…It might be inferred that the luminiferous æther, pervading all space, is not only highly elastic, but absolutely solid!!!".[77]

But the growing number of supporters of the wave theory chose to overlook the obvious problem that this posed for the motion of solid bodies

---

[77] Young, T. (1855), p 415.

such as planets through a rigid æther because the predictions of Fresnel's mathematical wave theory agreed so well with experiment. In 1821 Fresnel finally committed himself publicly to the idea that light is a transverse vibration: ordinary light, he said, consists of "…waves polarised in all directions." Arago, up to this point Fresnel's staunchest ally, quailed at the thought of light as a transverse wave and jumped ship. But by then his support was no longer necessary for Fresnel had at last established his reputation as France's leading authority on theoretical optics. As for questions about the æther, those weren't answered until 1905, as we shall see in the next chapter.

Despite Fresnel's success in accounting for every optical phenomenon then known, several leading Newtonians refused to run up a white flag. But the tide was turning in Fresnel's favour, though he did not live to savour the final triumph. By 1830 a new generation of physicists on both sides of the Channel, attracted by the mathematical elegance and simplicity of his wave theory, had begun to turn their backs on the corpuscular theory, which they considered to be based on far too many ad hoc assumptions. Fittingly, it was a Frenchman who administered the *coup de grâce* to the corpuscular theory. In 1862, Jean Foucault (1819–1868) succeeded in measuring the speed of light in both water and air and found that light travels more slowly in water than in air. Even the most ardent Newtonian had now to admit defeat, for one of the central tenets of the corpuscular theory, one that had been insisted upon by both Descartes and Newton, was that light speeds up when it enters a transparent medium.

Although Young is rightly given the credit for his discovery of the principle of interference and for his experimental demonstrations of the wavelike nature of light, without Fresnel's mathematical underpinning, the wave theory would not have replaced the corpuscular theory when it did. Of course, given the shortcomings of the corpuscular theory, sooner or later, another Fresnel would have come onto the scene. But when one reviews the situation as it was during the first couple of decades of the nineteenth century, it is abundantly clear that Fresnel had no peers. Apart from Young and Arago, neither of whom possessed the requisite mathematical ability to develop the mathematical underpinning that a solidly grounded wave theory of light required, no other scientists were prepared to countenance that it offered an alternative to Newton's corpuscular account of light. There was mathematical talent in abundance among physicists in France, and later in Britain, but in France that talent was used to develop ever more ad hoc explanations of optical phenomena based on the corpuscular theory.

Despite their cordial relations, Young always considered the wave theory as his brainchild and maintained that Fresnel had merely extended his ideas. Late in 1824, an exasperated Fresnel felt it necessary to put him right.

> I am far from laying claim to what belongs to you, Monsieur, as you have seen in the Supplement to the French translation of Thomson's Chemistry, as you will see also in the article I have just prepared for the European Review. I have declared with sufficiently good grace before the public, on several occasions, the priority of your discoveries, your observations and even your hypotheses. However, between ourselves, I am not persuaded of the justice of the remark in which you would compare yourself to a tree and me to the apple which the tree has produced; I am personally convinced that the apple would have appeared without the tree, for the first explanations which occurred to me of the phenomena of diffraction and of the coloured rings, of the laws of reflection and of refraction, I have drawn from my own resources, without having read either your work or that of Huyghens.[78]

Fresnel was quite right; at best, Young was John the Baptist to Fresnel's Christ. But it certainly didn't hurt Fresnel's cause to have Young on board for it added weight to his discoveries and theories.

By then Young had long since moved on to other matters. In 1811 he had been appointed as a physician at St. George's Hospital near Hyde Park in London. And in 1814 he had turned his attention back to languages when he began work on deciphering Egyptian hieroglyphs. Interest in these had been revived by the discovery of the Rosetta Stone in 1799 by French soldiers during Napoleon's Egyptian campaign. However, the stone ended up in London rather than Paris, having been surrendered to the British following their defeat of French forces in Egypt in 1801. At the time it was assumed that all hieroglyphs were symbols. But using the bilingual text inscribed on the Rosetta Stone, Young realised that this was not entirely correct: hieroglyphs have a phonetic as well as symbolic value; they stand for sounds as well as ideas.[79]

Curiously, he had a French rival in this, as he had had in optics, Jean-François Champollion (1790–1832). Although it was Champollion who succeeded in fully deciphering hieroglyphs, it appears that his early work was influenced by Young, something that Champollion was never prepared to acknowledge. As a result, relations between Young and Champollion were never cordial, as they had been between Young and Fresnel.

---

[78] Young, T. (1855), pp 401–2.

[79] The text inscribed on the stone is in three scripts, one of which is in hieroglyphs and another in Greek. The gist of it was that priests were exempted from having to pay tax. Plus ça change!

Both Fresnel and Young died prematurely, Fresnel from tuberculosis in 1827 at the age of 39 and Young a couple of years later from the effects of a diseased heart, aged 59. He was still at work on a dictionary of Egyptian hieroglyphs on the day he died. To a friend who suggested that in his condition he should take it easy, Young replied that it was "a great satisfaction to him never to have spent an idle day in his life".[80] It would have made a fitting epitaph for his tombstone.

If the British scientific establishment was largely indifferent to Young's principle of interference and his account of light as a wave, his explanation for Newton's Rings and supernumerary arcs was no sooner published than forgotten. His explanation of supernumerary arcs, in particular, was little known during his lifetime.

Several years after Young's death, Richard Potter (1799–1886), at the time a mature student at Cambridge University, concluded, in complete ignorance of Young's work on light, that a thorough explanation of the rainbow must be based on the principle of interference. He pointed out that Newton had offered no explanation for supernumerary arcs or for the fact that the colours in a rainbow depend upon the size of the drops. How else to explain the lack of colour in a fogbow, a rainbow formed in very small drops? As a committed supporter of the corpuscular theory, however, Potter faced a problem in reconciling corpuscular theory with the principle of interference. He overcame this by assuming that corpuscles are emitted from a source of light in sheets, which he called luminiferous surfaces, at regular intervals determined by colour. Mathematically, these luminiferous surfaces were identical to Huygens' wavefronts.

By tackling the problem of the rainbow in terms of wavefronts rather than individual waves as Young had, Potter's explanation for the rainbow was a huge step forward. It enabled him to show that a rainbow is the result of a wavefront that folds over on itself when it is reflected within the drop. According to Potter, on entering a drop a wavefront is in effect spilt in two: one due to light that enters the drop above the Cartesian rainbow ray and the other that enters the drop below the rainbow ray. Potter's mathematical analysis of the passage of the split wavefront through the drop shows that it emerges from the drop as two wavefronts that differ slightly in curvature and which intersect at a point that coincides with that of the Cartesian rainbow ray. The point of intersection, or cusp, which Potter identified as a caustic surface, is where the rainbow is brightest. But successive wavefronts also intersect at other points within the cone of the rainbow, which means that in

---

[80] Peacock, G. (1855), p 480.

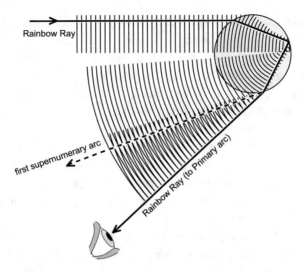

Fig. 8.6 Potter's explanation for the source of supernumerary arcs. The colours seen in these arcs is due to a combination of constructive and destructive interference between wavefronts emerging from a drop

monochromatic light a rainbow would be a series of bright and dark bands that grow fainter and narrower towards the centre of the rainbow. However, the range of colours (i.e. wavelengths) in sunlight means that in practice a rainbow consists a superposition of an indefinite number bows of different colours and spacings. This is why the dark bands of a monochromatic bow are never seen in nature: their place is taken by the bright bands of other colours. The result is a series of supernumerary arcs composed of mixtures of spectral colours (Fig. 8.6).[81]

Potter's account also showed how the width of the bright and dark bands is determined by the size of the drop: the smaller the drop, the wider the bands. As a result, when drops are very small a rainbow becomes all but colourless because the bands of each colour broaden to the point where they overlap one another completely to produce white light. Potter's account of the rainbow was published in 1838.[82]

Although Potter had explained why bows that are formed in tiny drops are colourless, he had overlooked an important feature of fogbows: they are noticeably smaller and have a broader arc than rainbows. Nor could he have accounted for this difference because his explanation was based on a theory of light that made no allowance for the effects of diffraction. His loyalty to the

---

[81] Potter, R. (1856), pp 78–87.
[82] Potter, R. (1838).

Fig. 8.7 A 360° fogbow. As with every rainbow, the photographer's head marks the bow's centre[83]

corpuscular theory prevented him from joining forces the majority of British scientists who had by then embraced Fresnel's wave theory of light. The application of this theory to the rainbow required more than an acceptance that light is a wave, however, it also called for formidable mathematical powers (Fig. 8.7).

One of the earliest British converts to Fresnel's views on light was George Biddell Airy (1801–1892), an astronomer and mathematician. Born in Alnwick, Northumberland in 1801, he went up to Cambridge in 1819 where he impressed his tutors with his mathematical abilities. He graduated Senior Wrangler—i.e. top in his year—in the Mathematical Tripos of 1822 and the following year won the Smith's Prize, the top mathematical prize awarded by the University. He was appointed Lucasian Professor shortly before his 26th birthday, though he soon moved on to other posts in search of a salary sufficient to support a wife and their growing family. Within ten years he was installed as Astronomer Royal, the post in which he spent the remainder of his professional life. Although most of Airy's energies went into astronomy, he also made important contributions in optics. One of these was to apply the mathematics of Fresnel's wave theory to the problem of the rainbow assuming, as Potter had, that a rainbow is the result of intersecting wavefronts.

---

[83] Photo by Brocken Inaglory, Wikimedia Commmons. https://tinyurl.com/4b4dmk59.

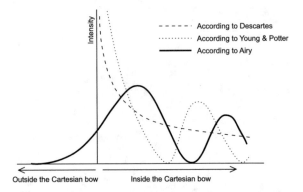

Fig. 8.8 A comparison of the variation in brightness across the arc of the primary rainbow according to Descartes, Young and Airy

In 1836, a few months before he left Cambridge for Greenwich to take up the post of Astronomer Royal, Airy employed Fresnel's wave theory to derive a complicated mathematical formula that gave the distribution and intensity of light emerging from a drop of water, a formula that has since become known as "Airy's rainbow integral" . In fact, deriving the rainbow integral was the least of Airy's achievement for its numerical solution proved almost as demanding as the derivation of the formula itself. Nevertheless, he succeeded in solving it and thereby established that according to Fresnel's wave theory of light a rainbow should be broader than that of either Newton or Young, and that the size of raindrops determines not only the spacing and colours of supernumeraries but also the radius of the bow. Probably the most unexpected outcome was the discovery that brightest part of the arc does not occur at the outer edge of the classic Cartesian rainbow, it actually falls slightly inside it, just beyond the notional rainbow ray (Fig. 8.8). Airy's account was published a couple of years later in the same volume of the Transactions of the Cambridge Philosophical Society that which carried Potter's paper.[84]

Airy was not an experimental scientist and left it to others to confirm his theoretical predictions. The first person to do this was William Hallowes Miller (1801–1880), a professor of mineralogy at Cambridge, who put Airy's theory to the test under laboratory conditions. In order to simplify things, Miller used a narrow, cylindrical stream of water in place of a drop, thus reducing the problem of observation from three dimensions to two. By this means, the resulting bows were reduced to a series of light and dark bands parallel to the stream of water rather than the curved arcs that would have

---

[84] Airy, G.B. (1838).

been observed in drops. By varying the diameter of the stream of water from approximately half a millimetre to as little as a third of a millimetre and illuminating it with light of different colours, Miller obtained experimental results that were broadly in line with Airy's theoretical predictions.[85]

But a wider acceptance of Airy's theory of the rainbow was hampered by its mathematical complexity. He had only been able to carry out calculations that established the position and relative brightness of first two bright bands. When improved methods of computation were used to find the position of further bands, discrepancies between theory and experimental results obtained under laboratory conditions surfaced. It became apparent that far from being the last word on the subject, Airy's account of the rainbow was but an approximation, albeit of a high order.

In any case, the theory had been tested using only monochromatic light. What should a natural rainbow look like according to Airy's theory? A natural rainbow is, of course, formed in a rain shower that contains a vast number of drops of different sizes illuminated by sunlight and seen against a background that may range from dark clouds to a clear blue sky. The Herculean task of working out the appearance of such a bow was taken on by Joseph Pernter (1848–1908), a professor of physics and meteorology at the University of Vienna. Pernter painstakingly calculated the position of Airy's multiple bows for several colours and drop sizes. Having done this he then added the results to find out the overall effect on the appearance of the bow to the eye by combining the resulting bands of different colours and widths.[86] From a theoretical point of view, the result is a useful rather than an exact guide to what one should see in rainbows formed in drops of different sizes.[87] These days a far more accurate simulation of rainbows formed by drops of different sizes can be accomplished in minutes using software based on a theory of the nature of light that was in its infancy during the period that Pertner was striving to close the gap between the abstractions of Airy's rainbow integral and natural rainbows.[88] Yet for a number of reasons unconnected with the limitations of Airy's theory, a modern version of Pertner's exercise gives perfectly satisfactory results as far as the eye is concerned.

---

[85] Miller, W.H. (1841).

[86] Pernter, J. M., Exner, F. M. (1922), p 565–88. For a summary of Pernter's work see Hammer, D. (1903).

[87] Tricker, R.A.R. (1970), pp 183–90.

[88] See: https://tinyurl.com/2sa2p5n4. Accessed 21/08/22.

9

The Electric Rainbow

> The whole domain of Optics is now annexed to Electricity, which has thus become an imperial science.[1]

The reason why Airy's rainbow integral fails to yield a completely accurate representation of the rainbow is that it is based on the idea that light is a mechanical vibration of a physical medium. But in 1821, the same year that Fresnel announced that light is a transverse wave travelling through an æther, Michael Faraday (1791–1867), who had been taken on as a laboratory assistant by Humphry Davy in 1812, had embarked on a program of research into electricity and magnetism that by the end of that century would result in a far more comprehensive account of the nature of light than was available to Airy, and to a belated realisation that there is more to the rainbow than meets the eye.

Faraday was the last person to make major discoveries in physics without any recourse to mathematics, and so was not himself the author of the new theory. During his lifetime he was deservedly hailed as the most successful and prolific experimental scientist of all time—a reputation that appears to have stood the test of time. But he was also a daringly imaginative thinker, arguably the most profound theoretician in the physical sciences of the nineteenth century, the first person to demonstrate experimentally the intimate connection between electricity, magnetism and light and grasp the implications of

---

[1] Lodge, O. (1907), p 289.

© The Author(s), under exclusive license to Springer Nature Switzerland AG 2023
J. Naylor, *The Riddle of the Rainbow*, Copernicus Books,
https://doi.org/10.1007/978-3-031-23908-3_9

this fact. Among the vast number of his discoveries two are particularly significant in the history of optics. In 1831 he found how to make electricity
from magnetism and in 1845 he demonstrated that magnetism affects the
polarisation of light. But his numerous discoveries were not in themselves the
revolution he wrought in physics; it was, as we shall see, the non-Newtonian
metaphysics that he developed in order to make sense of them.

None of this would have been possible, however, without Alessandro
Volta's electric battery. In the hands of men like Davy and Faraday, Volta's
invention came to rival the telescope and microscope as a catalyst for scientific progress for, like those instruments, it made possible major discoveries
that led to new ways of looking at the world.

It has been known since antiquity that a phenomenon akin to magnetism
can be produced by rubbing together two dissimilar materials such as glass
and wool or amber and linen, as long as everything is dry.[2] The phenomenon
is, of course, electricity and its similarity to magnetism is that electrically
charged bodies either attract or repel one another. But until the eighteenth
century, interest in electricity was at best marginal because it appeared to be a
trivial and somewhat feeble phenomenon, lightning notwithstanding. Indeed,
lightning was not recognised to be an electrical phenomenon until well into
the eighteenth century.

Sometime during the mid 1740s a Dutchman, Pieter van Musschenbroek
(1691–1761), Professor of Philosophy and Mathematics at Leiden University, inadvertently discovered a method for storing the electric charge created
when two unlike substances are rubbed together. During the course of an
experiment in which he was holding a glass flask filled with water that was
being electrically charged he received such a powerful shock from the flask
that for a few moments he could not breathe; he swore that "nothing in the
world would tempt him to try the thing over again...".[3] But he must have
had second thoughts for eventually he came up with a device that consisted
of a glass jar coated with metal sheath inside and out and which came to
be known as a Leiden Jar.[4] Although considerable amounts of electric charge
could be stored in a Leiden Jar, it proved difficult to control its output because
in use it produced only a brief and sometimes large surge of current. This was
enough for skilled and imaginative experimenters such as Benjamin Franklin
(1705–1790) to explore some of the properties of this mysterious fluid, as

---

[2] The materials must be electrical insulators to enable electric charge to accumulate on the surface of
the object.

[3] Nollet, J. A. (1746), p 1. English translation in Magie, W. M. (1963), pp 403–6.

[4] The modern version of the Leiden Jar is the *capacitor*, an important component in many electrical
apparatus.

electricity was believed to be, and in the process establish in 1752 that light-ning is a discharge of electrical energy identical to that produced by a Leiden Jar, though on a vastly greater scale.

In 1780, Luigi Galvani (1737–1798), a professor of anatomy at the University of Bologna noticed that the muscle of a dissected frog's leg would twitch when in contact with two dissimilar metals. He realised that this was an electrical phenomenon and concluded that electrical energy was somehow present in the fibres of the muscle. In his view, contact with metals merely served to draw that energy from the fibres.[5] Alessandro Volta, the leading authority on electrical phenomena at the University of Pavia, disagreed. He suggested that the current was due to contact between dissimilar metals and that the fluid in the frog's muscle provided a pathway for the current to flow between them. And to make his point he constructed the world's first electric battery in 1800 for the sole purpose of refuting Galvani, never imagining that the device would be of any further use.

Volta's battery consisted several of pairs of silver and zinc discs separated from one another by cardboard discs that had been soaked in brine and stacked one upon the other to create a short column. An electric current could be drawn from the device, which became known as a voltaic pile, by connecting the top and bottom discs of the stack with a wire. The current was not as large as that produced by the discharge of a Leiden Jar, but it was continuous and could be maintained until the chemical changes in the battery that are the source of the current cease to occur. But because the device had been designed in answer to Galvani's speculations about animal electricity, Volta could think of no further use for it than to employ it to study the physiological effects of an electric current by attaching the battery to different parts of his body. The culmination of these experiments involved inserting wires from one of his most powerful voltaic piles into his ears: "I received a shock in the head, and some moments after … I began to hear a sound … it was a kind of crackling with shocks, as if some paste or tenacious matter had been boiling… The disagreeable sensation, and which I appre-hended might be dangerous, of the shock in the brain, prevented me from repeating this experiment."[6] The sounds were probably due to earwax heated by the passage of current. He announced his invention and the results of this research in June 1800, in a letter written in French and addressed to the Pres-ident of the Royal Society, Sir Joseph Banks. A translation was duly published in the *Philosophical Transactions*.[7]

---

[5] Galvani, L. (1791). English translation in Magie, W.M. (1963), pp 420–7.

[6] Volta, A. (1800), pp 403–31. English translation in Magie, W.M. (1963), pp 427–31.

[7] Volta, A. (1800), pp 403–431.

Although Volta had communicated his invention in the pages of the *Philosophical Transactions*, when it came to demonstrating his device he judged that Paris was preferable to London. The French had recently driven the Austrians out of northern Italy, and were now its masters. Volta, quite reasonably, believed it was in his interest to ingratiate himself with the new authorities and so he travelled to Paris in September 1801 in order to demonstrate his invention to the members of l'Institut.[8] Among the audience was the republic's First Consul, Napoleon Bonaparte, then a member of l'Institut, who was heard to remark to one of the chemists present: "These phenomena belong to chemistry even more than to physics, and you ought to get hold of them."[9] At Bonaparte's urging, was awarded Volta a gold medal and a prize of 6000 Francs by l'Institut.

But if Volta could think of no better use for his battery than to employ it to study the effect of electricity on living creatures, others were quick to realise that access to a source of electricity that could be sustained for long periods opened up new and exciting avenues of research. As we saw in the last chapter, one of the first chemists to recognise the value of Volta's invention as a research tool was Humphry Davy. His background as a chemist gave him an insight denied to Volta into the process responsible for the electric current created by a battery. Volta believed that electricity was created through direct contact between unlike metals; in his view, the brine-soaked discs merely facilitated the passage of a current. But the fact that the metal electrodes of Volta's battery gradually eroded during use led Davy to conclude that the electrical current was due to a chemical reaction between the materials of which it was made. More importantly, if the current was due to a chemical reaction the reverse should be possible: the passage of a current through a substance should initiate reactions that might decompose it into its constituents. He declared that Volta's invention was "an alarm bell to experimenters in every part of Europe".[10] His hunch proved right, and, as we have already seen, he went on to discover several new elements by decomposing a variety of molten substances with powerful electric currents.

Although Faraday embarked on his scientific career as Davy's protégé, and had had to overcome the handicap of a humble and impoverished background by his own efforts as had Davy, the two men were poles apart in character. Where Davy brazenly proclaimed himself a genius and became increasingly and tiresomely socially ambitious with age, Faraday was always indifferent to status. He refused the highest honours that Britain has to confer

---

[8] Volta's letter was read at a meeting of the Royal Society on 7th November, 1801.

[9] Sarton, G. (1931), p 127.

[10] Cunningham, A., Jardine, N. (eds) (1990), p 222.

on her scientists: he turned down a knighthood, refused a university professorship specially created for him and spurned the offer of the presidencies
of both the Royal Society and the Royal Institution. He could have earned
a fortune carrying out research on behalf of any number of industries or
patenting his many discoveries, but chose to live simply with his wife in a
suite of rooms in the Royal Institution where, as its Director of Laboratory,
he was free to pursue his researches.

He had, by all accounts, the most engaging personality: straightforward,
modest, empathetic, warm, gentle, generous, all of which make him among
the most attractive figures in the history of science; hardly any of his contemporaries spoke ill of him. That such an apparently simple soul, bereft of
any mathematical aptitude, single-handedly came close to bringing Newtonian mechanical philosophy to its knees—had his peers been more attentive
and open to his ideas—seems hardly possible until one learns that there
was another aspect to his personality. As John Tyndall, his friend and close
colleague pointed out:

> We have heard much of Faraday's gentleness and sweetness and tenderness. It
> is all true, but it is very incomplete. You cannot resolve a powerful nature into
> these elements, and Faraday's character would have been less admirable than
> it was had it not embraced forces and tendencies to which the silky adjectives
> "gentle" and "tender" would by no means apply. Underneath his sweetness and
> gentleness was the heat of a volcano. He was a man of excitable and fiery
> nature; but through high self-discipline he had converted the fire into a central
> glow and motive power of life, instead of permitting it to waste itself in useless
> passion.[11]

That self-discipline owed much to two things: his religion and his wife,
Sarah. Both he and Sarah were Sandemanians, a tiny and exclusive Christian
sect based on a literal interpretation of the Bible. But while Sandemanians
were neither fire-and-brimstone nor sackcloth-and-ashes Christians, they did
demand strict adherence to their principles. They believed that the true
meaning of the Bible is to be found in the example of Christ's love and
self-sacrifice. That and the self-contained serenity of this tiny congregation
was almost certainly the source of Faraday's phenomenal capacity to devote
himself single-mindedly and selflessly to his work with little thought to material reward, which was in turn one of the keys to his exceptional success as a
scientist. Sandemanian doctrine was also a source of ideas that were central
to his view of the world and which guided his research: that God created a

---

[11] Tyndall, J. (1868), p 37.

Universe in which nothing goes to waste (which plausibly may be construed as a common sense version of the principle of the conservation of energy) and in which there is an underlying unity. In other words, all phenomena are somehow interconnected.

Sarah was also essential to his success. She took no interest in his scientific work but provided him with a warm and peaceful home, a respite from his hours in the laboratory. She was, he once told her, "...a pillow to my mind...".[12] Their union was a love match that endured till death; there can't be many scientists who interrupt a letter to their wife describing the scientific work they are engaged upon by declaring "I am tired of the dull detail of things, and want to talk of love to you".[13] There were no children, but plenty of nephews and nieces. During their frequent extended visits, Faraday would sometimes entertain them in his laboratory with impromptu experiments and construct simple scientific toys for their amusement. At other times, he put his laboratory to domestic use as the need arose, manufacturing household chemicals for Sarah and once a year brewing a large quantity of ginger wine (his young visitors, however, preferred shop-bought sweets to the ones made in the laboratory by their uncle). He was very far from being the caricature detached, unemotional scientist; according to Tyndall "There was no trace of asceticism in his nature. He preferred the meat and wine of life to its locusts and wild honey."[14]

Faraday's father was a jobbing blacksmith who struggled to provide for his large family; the son's prospects were correspondingly poor. His formal education consisted of "...little more than the rudiments of reading, writing and arithmetic."[15] The best he could expect was to take up a trade. In 1804, when he was 13, he went to work for a bookseller to whom he was apprenticed as a bookbinder. It proved to be a good move, for as he later told a friend: "There were plenty of books, and I read them." He was especially drawn to those on electricity and chemistry.

For better or worse, like every other budding scientist of humble origins in the early decades of nineteenth century Britain, Faraday had to raise himself by his own bootstraps. There were no institutions of learning that could offer poor but promising young men a rigorous grounding in physics, chemistry and mathematics as there was in France. Extraordinarily, unlike almost every major city on the Continent, London had no university of its own. Indeed there were only two universities in the whole of England at the time: Oxford

[12] Letter to Sarah Faraday 14th August, 1863. In: James, F.A.J.L. (2011).

[13] Bence Jones, H. (1870), Vol 1, pp 364–5.

[14] Tyndall, J. (1868), p 151.

[15] Bence Jones, H. (1870), Vol 1, p 7.

and Cambridge—Scotland had four: St Andrews, Glasgow, Aberdeen and Edinburgh. Oxford and Cambridge did everything in their power to prevent rival universities being set up in English cities. So the best that was on offer to anyone unable by birth or circumstance to attend these two venerable institutions was a scattering of societies for amateur scientists and the public lectures on scientific subjects such as those that Faraday attended in his spare time. His initial scientific education was thus largely in his own hands. Fortunately, young Faraday was determined on self-improvement and took every opportunity on offer. As well as attending public lectures, he took drawing lessons from a fellow lodger and arranged for lessons in English to improve his grasp of the language.

It seems that his interest in science was first aroused by an article on electricity in a copy of the Encyclopaedia Britannica that he was required to bind. But the book that inspired him most was *Conversations on Chemistry* by Mrs. Jane Marcet (1769–1858), from which he learned about Humphry Davy's ideas on the nature of electricity and its uses in chemistry. Jane Marcet had no formal training in chemistry but had become interested in the subject as a result of attending Davy's lectures and aimed her book at Davy's female audience.[16] It took the form a dialogue between a knowledgeable Mrs. B and pair of earnest young women, Emily and Caroline, and in addition to explaining chemistry, emphasised the importance of experiments as an aid to understanding. It inspired Faraday to try his hand at a few simple experiments in a backroom of the bookseller's shop using whatever chemicals and vessels that were at hand.

Yet Faraday was no prodigy. In a letter to a friend on the occasion of the death of Mrs. Marcet he wrote:

Do not suppose that I was a very deep thinker, or was marked as a precocious person. I was a very lively imaginative person, and could believe in the *Arabian Nights* as easily as in the Encyclopaedia. But facts were important to me, and saved me. I could trust a fact, and always cross-examined an assertion. So when I questioned Mrs Marcet's book by such little experiments as I could find means to perform, and found it true to the facts as I could understand them, I felt that I had got hold of an anchor in chemical knowledge, and clung fast to it.[17]

---

[16] Jane Marcet was a populariser, not a professional scientist. Nevertheless, her "Conversations on Chemistry", published in anonymously 1806, became a bestseller. Her husband was a Swiss doctor.

[17] Bence Jones, H. (1870), Vol 2, p 401.

Faraday was therefore already well acquainted with Davy's ideas when he was given a ticket in the spring of 1812 by one of his employer's customers for Davy's final series of lectures at the Royal Institution in which he demonstrated that Lavoisier was mistaken in maintaining that oxygen was the essential component of all acids. This was also the year in which Faraday's apprenticeship was drawing to a close. Desperate to escape a future as a bookbinder, which he believed would rule out any chance a career in science, Faraday wrote to Sir Joseph Banks to ask for a job, any job, in a scientific capacity. Banks didn't reply. But a stroke of luck saved Faraday from the trade for which he seemed destined. Davy had injured his eye in an explosion and needed someone to keep his laboratory notes while he recovered. Faraday had made a written record of Davy's lectures that he had attended earlier that year and had shown them to the person who had given him the lecture ticket. And it was this person who recommended Faraday to Davy. Although Davy engaged Faraday for only a few days on that occasion, he hired him on a permanent basis a few months later to replace an assistant who had been dismissed from Royal Institution for brawling in one of its laboratories.

Within weeks of his appointment, Faraday was given further responsibilities: alongside his duties as a laboratory assistant he was asked to help out during lectures. It wasn't long before Davy recognised his assistant's potential and in mid 1813, barely a year after hiring him, he invited Faraday to accompany him on an extended tour of France and Italy. Davy was then at the height of his fame as Britain's leading chemist, and had been awarded the Prix Napoléon a few years earlier by l'Institut for his work in electricity. This was the prize that had been established at Napoleon's bidding in recognition of Volta's invention and which henceforth was to be awarded to "the person who, by his experiments and discoveries, shall, according to the opinion of [l'Institut], advance the knowledge of Electricity and Galvanism as much as Franklin and Volta did."[18]

The tour took place in the most extraordinary circumstances because France and Britain were at war and a British Army under the Duke of Wellington was about to cross the Pyrenees into France. Despite the fact that the only Britons then in France were either held prisoners or under house arrest, Napoleon saw some political advantage in allowing Britain's most distinguished chemist to come to France. So when Davy applied to visit France to collect the prize he was granted a passport for himself and his party, something for which he was heavily criticised in the British press.

---

[18] Paris, J.A. (1831), p 259.

Davy's defence was that "if the two countries or governments are at war, the men of science are not."[19]

The third member of the party was Davy's wife. Davy had married Jane Apreece (1780–1855), a wealthy heiress and a widow, on 11th April, 1812, three days after he was knighted for his services to chemistry. As the tour wore on, Lady Davy began to go out of her way to put Faraday in his place, though to judge from his diary entries he gave as good as he got. The marriage did not flourish; a close friend of both Davys, the Scottish novelist and poet, Walter Scott (1771–1832), noted in his journal "She has a temper and Davy has a temper, and these two tempers are not one temper, and they quarrel like cat and dog, which may be good for stirring up the stagnation of domestic life, but they let the world see it, and that is not well".[20] Within a few years Sir Humphry and Lady Davy were leading largely separate lives, though they never divorced. Davy, an enthusiastic fisherman, consoled himself with trips to lakes and streams of Britain, solo tours of the Continent and writing poems. In spite of their estrangement, Jane travelled to Rome in May 1829 to be with Davy during his final days; he died later that month in Geneva from a stroke, nineteen days after the death of Thomas Young.

While in Paris, Davy and Faraday dined with Count Rumford, whose disastrous marriage to Marie Lavoisier (1758–1836), the widow of Antoine Lavoisier, might have served as a belated warning to Davy. Believing that he was unappreciated in Britain, Rumford had gone to live in France where he had met and married the widow Lavoisier. Antoine Lavoisier was France's greatest chemist, but he earned his living as a tax collector for the *Ancien Régime* and was duly guillotined during the Reign of Terror.[21] Rumford and Madame Lavoisier de Rumford (as she insisted that she be known to preserve the great man's name, much to Rumford annoyance) discovered within months of tying the knot that they were chalk and cheese; according to Rumford's daughter, "he was fond of experiments, and she of company".[22] Moreover, both were strong, wilful, independent characters.

In fact, Marie was well versed in the sciences having assisted her first husband in his laboratory, dealt with his correspondence, learned English so that she could translate chemical texts from English into French and drawn all

---

[19] Paris, J.A. (1831), p 261.

[20] Scott, W. (1891), pp 108–9.

[21] Joseph-Louis Lagrange (1736–1813), an Italian mathematician and astronomer, who was a member of the *Académie des Sciences*, lamented that "It took them only an instant to cut off that head, but France may not produce another like it in a century". Lavoisier's father-in-law, also a tax collector, was guillotined on the same day.

[22] Ellis, G.E. (1871), p 602. Rumford wrote a letter to his daughter on every anniversary of his wedding to Marie, giving an account of a year's worth of marital strife.

the figures for the *Traité Élémentaire de Chimie*, his most important work.[23] But, evidently, her appetite for science had waned by the time Rumford appeared on the scene, and very soon after getting married found that they had nothing in common. Things came to a head when Madame Lavoisier invited a large group of her friends to their Paris house knowing that they would annoy Rumford. Rumford refused them entry and his wife, who knew of his fondness for flowers, retaliated by pouring several kettles of boiling water over his prized flowerbeds.[24] It led to their separation a few months later and a modest improvement in their relationship; there was no divorce. Rumford died a few months after Davy's visit.

It has been said that no experience was ever lost on Faraday. He made the most of his time on the continent: he visited museums, factories, met some of the leading scientists of the day including Arago and André-Marie Ampère (1775–1836), assisted Davy in his researches and learned some French.[25] For his part, Davy put his hosts' noses out by usurping the work of France's leading chemists on the question of whether a recently discovered purple-coloured substance was an element or a compound. While in Paris, Davy was surreptitiously provided with a sample of the substance by Ampère and, using apparatus that he had brought with him from England, quickly established that it must be an element and even coined a name for it: iodine. His hosts were not pleased—nor were they minded to accept his findings.[26] Nevertheless, throughout the visit, the French savants were unfailingly courteous and helpful to Davy despite the fact that he made little effort to return the compliment, in part because he did not wish to be seen to as being over friendly with the enemy.

Months later, in Florence, Davy employed a large convex lens that belonged to the Grand Duke to set diamonds alight.[27] The purpose of this seemingly whimsical experiment was to determine the chemical nature of diamond.[28] The suspicion was that it was composed of carbon, and if that were so then burning a diamond in pure oxygen should produce only carbon dioxide. A diamond was placed in a platinum crucible within a glass globe filled with oxygen and the sun's rays concentrated on it using the Duke's lens.

---

[23] Lavoisier, A. (1789).

[24] Ellis, G.E. (1871), p 560.

[25] Faraday's command of French was very good and he was also familiar with Italian, but never learned German, something that he regretted.

[26] It is more than likely that Ampère gave Davy the sample to spite the *Académie's* Laplacian faction. His fellow academicians were furious with him.

[27] The lens was some 35 cm in diameter and was used in conjunction with a smaller 7.5 cm diameter convex lens.

[28] Lavoisier had performed the same experiment in 1772.

Faraday noted in his journal that when it eventually ignited after some 45 min "The diamond glowed brilliantly with a scarlet light inclining to purple, and when placed in the dark continued to burn for about four minutes."[29] The product of the combustion was found to be carbon dioxide. Davy concluded that difference in appearance between diamond and other forms of carbon such as graphite and charcoal lay solely in the arrangement of carbon atoms in each of these materials.

The tour was cut short in the spring of 1815 when they learned that Napoleon had escaped from Elba. It is, of course, mere coincidence that the careers of the two men who were to usher in a new era in physics by providing alternatives to the prevailing Newtonian orthodoxy were directly affected by this event, but perhaps one worth mentioning, given the immediate effect it had on the lives of both men. In Fresnel's case, as we saw in the last chapter, it resulted in a temporary suspension from duty during which he began the series of experiments that led him to the wave theory of light; in Faraday's, it took him back to the laboratory in the Royal Institution where he eventually came up with an alternative to Newtonian action-at-a-distance and thereby provided a new framework for Fresnel's light waves.

On his return to the Royal Institution, Faraday was given a promotion and from 1818 to 1830 worked primarily as an analytical chemist. This period was in all but name a second apprenticeship, this time as a chemist. Nevertheless, although he made several minor discoveries in chemistry—he isolated benzene and liquefied chlorine—he was to make his name for his research into electricity and magnetism, research that was prompted when he was asked to write an account for a journal of a sensational discovery by a Danish natural philosopher, Hans Christian Oersted (1777–1851).

Oersted was a leading exponent of *Naturphilosophie,* a diverse school of thought that was particularly influential in Germany and which was part of the widespread Romantic reaction to the perceived narrowness and materialism of the eighteenth century science and philosophy. *Naturphilosophs* were united in their rejection of Newtonian science and consequently spurned the mathematical physics of the French. Their opposition stemmed principally from a belief that nature is an organism rather than a mechanism and that the task of a natural philosopher is to discover its inner workings. Newton, as we have seen, had an aversion to speculation, at least as far as physics was concerned. However, although in the main *Naturphilosophie* hindered the progress of science in Germany during the first decades of the nineteenth century, one of its central tenets proved to be extremely fruitful. This was

---

[29] Faraday, M. (1991), pp 75–6.

that all forces in Nature—electricity, magnetism, gravity, light etc.—shared an underlying unity.[30] Moreover, their interaction was believed to be the manifestation of a struggle between polar opposites such as that between positive and negative charges in electricity, north and south poles in magnetism and brightness and darkness in light.[31]

Where electricity and magnetism are concerned, the idea that they are somehow linked wasn't such an outlandish idea because there was a great deal of anecdotal evidence for a connection between them. Sailors had long known that the polarity of a compass needle is sometimes reversed when a ship is struck by lightning.[32] And in 1731 a Yorkshire cutler had reported that a collection of knives and forks that had been struck by lightning had been magnetised when he noticed that they were able to attract iron nails.[33] The phenomenon could even be reproduced in the laboratory. Benjamin Franklin had managed to magnetise an iron needle by employing a Leiden Jar to pass a large current through it.[34] A few decades later, in 1801, a French chemist, Nicholas Gautherot (1753–1803), noticed that two parallel wires placed close together attracted one another when they were both connected in parallel to a battery.[35] But despite these hints, the hostility of the majority of French and British scientists towards *Naturphilosophie* prevented them following up such evidence. In one of his lectures at the Royal Institution Thomas Young, whose scientific insights were so often spot on, had insisted that "...there is no reason to imagine any immediate connexion between magnetism and electricity."[36] But in 1820 irrefutable evidence of such a link was found when Oersted at long last stumbled across an effect he had sought for a decade.

Oersted had been struck by the fact that when electricity flows through a narrow wire it gets hot and, if the wire is very narrow, glows faintly. He surmised that in an even narrower wire electricity might be converted into some other form of energy, perhaps magnetism. Confirmation of his hunch occurred towards the end of a lecture he gave on the nature of electricity and magnetism. The necessary apparatus, which consisted of a tiny magnetic compass placed slightly below to a short length of very thin platinum wire, was set up in preparation for his lecture but he didn't carry out a trial run to

---

[30] Humphrey Davy was influenced by *Naturphilosphie*, which he learned from his friend, the poet Samuel Taylor Coleridge (1772–1834).

[31] Gravity proved difficult to fit into this scheme, and indeed remains something of the odd man out in the unifying theories of physics to this day. Gravity is always a force of attraction.

[32] Anon (1676), pp 647–8.

[33] Dod,P. (1735), pp 74–5.

[34] Franklin, B. (1941), pp 242–3.

[35] Gautherot, N. (1801), p 209.

[36] Young, T. (1845), Vol 1, p 538.

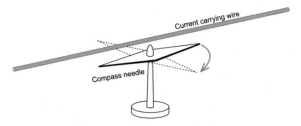

Fig. 9.1 Oersted's experiment demonstrating the magnetic effect of an electric current. When a current flows through the wire the compass needle swings in the direction shown by the arrow and sets itself at right angles to the wire

check that everything was working—he was famously ham-fisted, and usually relied on an assistant to set up his experiments.[37] To his surprise, when he got around to connecting the wire to a battery, the compass needle stirred. But "these experiments were made with a feeble apparatus, and were not, therefore, sufficiently conclusive, considering the importance of the subject…"[38] Given that he had been searching for such an effect for several years, it is astonishing that he didn't get around to repeating the experiment for another 3 months. This time he employed a more powerful battery and having satisfied himself that the effect was reproducible he dashed off a brief account of his discovery in Latin and sent copies to the leading scientists of Europe.[39] It was the last time that Latin was used to communicate a major scientific discovery (Fig. 9.1).[40]

That an electric current has a magnetic effect wasn't the only surprise: the real eye-opener for Oersted was that it caused a compass needle to swivel until it set itself obliquely to the direction in which the current flowed through the wire. Hitherto, it had been assumed that all forces acted either directly towards or away from their source, as they do in the case of Newtonian gravitational attraction between bodies.[41] But here the force exerted by an electric current on a magnet appeared to act sideways.

A couple of months later, on learning of Oersted's discovery and seeing the experiment performed by Arago at a meeting of the Académie des Sciences, Ampère, who hitherto had shown little interest in either electricity or magnetism, leapt into action.[42] He realised that the reason that

---

[37] C. Hansteen in a letter to M. Faraday. In: Bence Jones, H. (1870), Vol 2, p 390.
[38] Oersted, H.C. (1820), p 273.
[39] Christensen, D.C. (2013), p 347.
[40] For a facsimile of Oersted's original paper see: Sarton, G. (1928).
[41] Such forces were known as *central forces.*
[42] Williams, L. P. (1983).

the compass needle came to rest obliquely to the wire was that Oersted had made no allowance for the effect of the earth's magnetic field. When Ampère repeated the experiment in such a way as to compensate for the earth's magnetic field, he found that that the compass needle always sets itself at right angles to a current-carrying wire. He then made a bold suggestion: magnetism is in fact a wholly electrical phenomenon, which, he said, implied that a permanent magnet is composed of a vast number tiny electric currents.[43] Moreover, if magnetism is an electric phenomenon, electric currents must act directly on one another—i.e., wires placed side by side must either attract or repel one another when currents flow through them. Experiments duly confirmed this, all of which enabled Ampère to develop a mathematical theory of the phenomenon based on Newtonian principles of action-at-a-distance, and which he named "electrodynamics".

No sooner did Davy receive the account a of Oersted's discovery than he hastened to the Royal Institution to share the news with Faraday and they immediately repeated the experiment. In the meantime, Wollaston had suggested that Oersted's discovery might be explained by assuming that an electric current spirals along a wire (rather than travelling directly through it) and concluded that a current-carrying wire should therefore spin about its axis in the presence of a permanent magnet. But all attempts to make this happen came to naught.

When Faraday got around to investigating the phenomenon more thoroughly, in preparation for the article he had been commissioned to write on the subject, he noticed that a permanent magnet causes a current-carrying wire to move sideways: "The effort of the wire is always to pass off at a right angle from the pole, indeed to go in a circle around it."[44] To demonstrate this, he constructed a device in which a wire suspended vertically above the end of a permanent magnet rotates about its pole when a current flows through the wire. When he published an account of these discoveries in October, 1821,[45] he found himself propelled to the front ranks of European science. He had, in fact, discovered the principle of the electric motor, though he never got around to designing and constructing such a device.[46]

But publication of these discoveries harmed his reputation for a time in Britain, for it was incorrectly assumed that he had made use of Wollaston's

---

[43] He was quite right: we now know that these electrical currents are due to the electrons that orbit the atoms.

[44] Faraday, M. (1932), pp 49–57.

[45] Faraday, M. (1822), pp 416–21.

[46] The first electric motor capable of turning machinery was invented by William Sturgeon (1783–1850) in 1832. But electric motors only became a practicable source of motive energy in the 1870s when electric dynamos replace batteries as a source of electric current.

ideas without acknowledgement. Indeed, it marked the beginning of the end of his close relationship with Davy, for Davy sided with Wollaston. So much so that a couple of years later, Davy, who had replaced Sir Joseph Banks as President in 1820, opposed Faraday's election as a Fellow of the Royal Society in 1823.

Oersted's discovery opened up another possibility: if electricity creates magnetism then the reverse should also be possible: magnetism creates electricity. But under what circumstances? During the decade following his discovery of the motor effect, Faraday made several unsuccessful attempts to use magnetism to create an electrical current. The problem was that he had not fully understood Oersted's experiment—nor had Oersted, though that is by-the-by. Faraday expected an electric current to be created in a wire simply by placing it in a magnetic field. What he had not appreciated was that in Oersted's experiment the electricity is not static: it is travelling through the wire. Placing a wire in a static magnetic field is therefore not a mirror image of Oersted's experiment.

Faraday discovered this vital fact on 29th August, 1831. His apparatus consisted of two separate coils of insulated copper wire wound on opposite sides of a ring of solid iron.[47] One of the coils created a magnetic field when connected to a battery and the other coil was hooked up to a galvanometer, a device for detecting electric current (Fig. 9.2).[48] As luck would have it, the galvanometer was already hooked up to the second coil before Faraday connected the first coil to a battery and so he was able to notice the tell-tale sign of an electric current in the second coil: a momentary flick of the galvanometer needle as the current surged through the first coil and the resulting magnetic field increased in strength. The needle gave another brief flick in the opposite direction when he disconnected the first coil from the battery.[49] But there was no sign of a current in the second coil as long as a steady current flowed through the first coil, i.e. as long as there was a steady magnetic field within the torus.

It was several weeks before Faraday had a thorough understanding of the underlying process: an electric current is created in a conductor such as a wire whenever there is relative motion between it and the magnetic field within which it is placed.[50] Seven weeks later, on 17th October, he produced

---

[47] He chose a ring because he considered that was the best shape to confine the magnetic effect produced by the coil of wire. The diameter of the ring was 15 cm and it was 2.5 cm thick.

[48] The version of galvanometer used by Faraday consist of a compass needle placed within a coil of wire. When a current flows through the wire, the needle is deflected. The device is, indeed, based on Oersted's experiment.

[49] Faraday had inadvertently invented the electrical transformer.

[50] Faraday called this process *induction*.

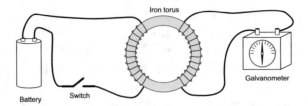

Fig. 9.2 Faraday's discovery of electromagnetic induction. When the switch is closed, an electric current flows through the left hand coil and induces a brief current in the right hand coil. The presence of the current is detected by the movement of the needle of the galvanometer

a current by thrusting a permanent magnet into a coil of wire. As with the iron ring, this current was short-lived, lasting only as long as the magnet was in motion, and it varied in strength and direction depending on the speed and direction of the magnet's motion. Ten days after this he constructed a primitive electric dynamo that produced a small but steady electric current. But as with his discovery of the motor effect, he didn't develop the device further.[51]

These discoveries confirmed Faraday's status as Europe's leading experimental scientist and set him off on a wide-ranging investigation into the relationship between electricity and matter. Over the following decade there were discoveries galore, too numerous and, for the most part, too technical to describe here. The upshot was that he fashioned new ways of understanding electrical phenomena and coined much of the vocabulary of modern electrical science. At the same time suspicions long held began to crystallise into a conviction: that electrical, magnetic and gravitational forces exist in the space between bodies rather than within the bodies themselves. He came to believe that space is not an inert void within which particles interact with one another either by direct contact or through the instantaneous action-at-a-distance of the Newtonians of the French school of mathematical physics. Instead, Faraday concluded, magnetism and electricity extend their influence throughout space and therefore space itself participates in all physical processes that involve force. As a corollary, he came to think of atoms not as material bodies (i.e., tiny, solid particles) but as nodes within a web of forces and that interactions between these immaterial atoms involve disturbances that travel at finite speed through the web. In other words, the forces that separated bodies exert on one another do not act instantaneously.

---

[51] The first dynamo based on Faraday's principles was built in 1832 by Hippolyte Pixii (1808–35), a French instrument maker.

Towards the end the 1830s, Faraday's health began to fail. This has been attributed to his long hours in his laboratory and to the sustained mental effort his ground-breaking work demanded of him as well as his exposure to toxic chemicals in the course of his research. He suffered from headaches, giddiness and, most alarmingly, memory loss. In 1843 he wrote to a friend that his memory was so poor "...that I cannot remember the beginning of a sentence to the end—hand disobedient to the will, that I cannot form the letters, bent with a certain crampness, so I hardly know whether I shall bring it to a close with consistency or not."[52] He abandoned his researches completely for two years. However, though he never returned to full health, by 1844 he had recovered sufficiently to resume work. Remarkably, this led to his greatest legacy: the idea that electrical, magnetic and gravitational forces associated with bodies exist in the space between bodies rather than in the bodies themselves.

Like Oersted, Faraday had long believed that there is an underlying unity of forces and that therefore every force should affect every other force in some manner: electricity affects magnetism and vice versa. But what of the effect of electricity on light? He began searching for such a link soon after his discovery of the motor effect in 1821, but failed to find one. But in September 1845, he succeeded with magnetism. Following a suggestion by William Thomson (1824–1907), one of the very few physicists of the time to take Faraday's ideas seriously, that a strong electric field ought to create a mechanical strain in glass that should affect the polarisation of light traversing the glass, Faraday again failed to detect such an effect.[53] But when he tried the experiment with magnetism he found that when glass is placed in a powerful magnetic field, the plane of polarisation of a beam of light is indeed altered.

Faraday's scepticism about the possibility of Newtonian action-at-a-distance sprang from his discovery that the material that occupies the space between electrically charged bodies or between magnets affects the force they exert on one another. Furthermore, it was obvious from his experiments that these forces don't act directly between bodies but curve out into the surrounding space as is so graphically illustrated by that stalwart of school physics experiments in which iron filings are sprinkled around a magnet to reveal the shape of its magnetic field. In fact, this is precisely how Faraday envisioned his immaterial web of forces: he believed that what he called "tubes of force" are real. Moreover, his experiments convinced him that magnetic and electric lines of force are in state of tension, rather like stretched elastic

---

[52] Faraday, M. (1899), p 109.

[53] This effect was discovered in 1875 by the Scottish physicist, John Kerr (1824–1907).

bands and that neighbouring parallel lines of force repel one another. Physicists still find Faraday's analogies a useful way to think about the properties of magnetic and electric fields.

At the same time, Faraday came to doubt the existence of the æther, the elephant in the room of Fresnel's wave theory of light. As we saw in the last chapter, Young had been the first to spot the difficulty: an æther with the properties necessary to transmit light as a mechanical vibration creates more problems than it solves. But Fresnel's mathematical theory was so successful in accounting for all the known properties of light that everyone was prepared to turn a blind eye to the incongruity of a medium rigid enough to transmit Fresnel's light waves but which offered no discernible resistance to the motion of material bodies. Faraday, to whom mathematics was a closed book, could take no comfort from Fresnel's equations; he was forced to confront the problematic nature of the æther head-on. Almost alone among his peers, he doubted that such a medium was feasible, even though he was aware that it might provide a mechanism for his own lines of force between bodies.[54]

A year later, in 1846, in order to fill time at the end of one of his lectures, he casually posed a rhetorical question and voiced an idea that had been at the back of his mind for more than a decade: might it be possible that "vibrations which in a certain theory [i.e. wave theory of light] are assumed to account for radiation and radiant phaenomena may … occur in the lines of force which connect particles, and consequently masses of matter together; a notion which as far as is admitted, will dispense with the æther, which in another view, is supposed to be the medium in which these vibrations take place."[55] In other words, Faraday was suggesting that light is what we now call an electromagnetic wave.

How did a mathematical illiterate outdo the best mathematical physicists in Europe at their own game? The fact is that Faraday was never the laboratory drudge that emerges from the way he is often portrayed. He was a natural philosopher in the fullest sense of that calling, perhaps one of the greatest there has ever been. What he sought was a deeper understanding of nature than met the experimenter's eye. Accordingly, his experiments almost always served a larger purpose and were seldom an end in themselves. He always took the broader view of things and over the years his discoveries in electricity and magnetism became stepping-stones to a non-Newtonian metaphysics. His success in this enterprise was due as much to his powerful and inventive imagination as to a remarkably open-minded attitude towards

---

[54] The only other scientist who agreed with Faraday on this point was the staunch Newtonian, Sir David Brewster.

[55] Faraday, M. (1846), p 345.

his own ideas. In the opinion of Tyndall, Faraday "...united vast strength with perfect flexibility. The intentness of his vision in any direction did not apparently diminish his power of perception in other directions; and when he attacked a subject, expecting results he had the faculty of keeping his mind alert, so that results different from those which he expected should not escape him through preoccupation."[56] So, where the uncompromising French savants were able to come up with clever mathematical explanations that papered over the ever-increasing number of cracks in their action-at-a-distance philosophy, Faraday's lack of mathematics forced him to search for alternative ideas to explain the results of his experiments.

Deteriorating health forced him to resign from the Royal Institution in 1862, after almost 50 years of service. By then he had become the Grand Old Man of British science. But his reputation rested on more than his laboratory discoveries than his theoretical ideas. He was also a successful populariser of science and a brilliant lecturer, easily a match for the more flamboyant Davy. One of his enduring legacies is the famous Christmas Lectures for children, which are still given yearly by leading scientists in the same lecture room at the Royal Institution that he had used. During his time at the Royal Institution, Faraday gave 19 of these lectures, and some were published in book form.[57] He and Sarah retired to a Grace-and-Favour house in Hampton Court provided by the Queen Victoria on the recommendation of her husband, Prince Albert. Faraday became increasingly senile and died on 25 August 1867.

Faraday's inability to express his ideas in the rigorous and precise language of mathematics meant that his scientific legacy might have been limited to his experimental discoveries, important though they are, had it not been for James Clerk Maxwell (1831–1879). Taking his cue from Faraday's ideas about lines of force, Maxwell, ranked by no lesser authority than Albert Einstein as second only to Newton in the pantheon of physicists, united Faraday's several discoveries and speculations in a grand mathematical theory of electromagnetism and in the process laid the foundations for much of the non-Newtonian physics of the twentieth century, including Einstein's own Theory of Relativity.

Maxwell was born in Edinburgh in 1831, a few months before Faraday discovered electromagnetic induction.[58] An only child—an older sister had died in infancy—he spent an idyllic childhood in bucolic isolation on the family estate in Galloway, some 90 miles south west of Edinburgh. Almost

---

[56] Tyndall, J. (1868), p 20.

[57] Faraday, M. (1908).

[58] 1831 was also the year in which Charles Darwin (1809–1882) embarked on HMS Beagle.

as soon as he could talk, he displayed that most important instinct of genius: never to take anything for granted. It manifested itself in a ferocious curiosity: nothing escaped his attention. His mother described him as a "…very happy man…he has great work with doors, locks, keys, etc. and 'Show me how it doos' is never out of his mouth."[59] He was fortunate in having parents willing and able to indulge his inexhaustible desire to know "What's the go of it?". And when an answer didn't satisfy him he would follow up with "But what's the particular go of it?"[60] He was especially close to his father, a halfhearted lawyer by profession and an enthusiastic amateur engineer by vocation, his mother having died from an abdominal cancer when Maxwell was 8 years old. The father's practical bent was particularly important to the boy's early intellectual development.

Maxwell's real talent was for mathematical physics, though that, of course, took several more years to materialise. Sent away to school in Edinburgh at the age of ten, he found himself the butt of his classmates' derision; they fastened on his country ways and tormented him unmercifully. Astonishingly, both they and his teachers took him for a bumpkin and nicknamed him "Dafty". But Maxwell was blessed with other gifts in addition to his intellect: he was unusually good natured and forgiving. As a consequence, he was able to endure the taunts and demonstrated his intellect by gaining school prizes in mathematics, English and divinity.

At the same time, his father's connections led to introductions to adults who could appreciate his potential and who, when he was only 14 years old, arranged for his first mathematical paper to be read at a meeting of the Royal Society of Edinburgh and published in its journal. This was on the mathematics of drawing ovals and although the subject had originally been tackled by Descartes, Maxwell's approach to the topic was considered to be superior.

Barely a year later he was attending the University of Edinburgh. Despite his obvious aptitude for science and mathematics, his father wanted him to study law. Even in mid-nineteenth century Britain, science was an uncertain profession, with few opportunities and poorly rewarded. But the father's friends, who had become aware of Maxwell's intellectual promise, prevailed on him to drop his objections.

Although Edinburgh provided a broad education in keeping with Scottish educational ideals—Maxwell studied philosophy and literature along with mathematics and science—Cambridge University had become an intellectual powerhouse of mathematical physics and was the obvious destination for any

---

[59] Campbell, L., Garnett, W. (1882), p 27.
[60] Campbell, L., Garnett, W. (1882), p 28.

Briton with serious scientific ambitions. Maxwell was advised to leave Edinburgh before graduating and travel south where, after a false start, he enrolled in 1850 in Newton's old college, Trinity. He found himself in his element: his intelligence, wit and eccentricity endeared him to his fellow undergraduates. Everyone who came across him realised that he was destined for great things and took the unconventional Scot in their stride.

While it is not necessary for brilliant and original people to be eccentric, many are, and Maxwell certainly had his fair share of mannerisms and idiosyncrasies. These were evident even in his youth when "His replies in ordinary conversation were indirect and enigmatical, often uttered with hesitation and in a monotonous key...".[61] Years later a Cambridge acquaintance recalled that "...his love of speaking in parables, combined with a certain obscurity of intonation, rendered it often difficult to seize his meaning".[62] Even at his most lucid, his quizzical turn of mind could make the most ordinary of conversations with him something of an ordeal.

His memory and powers of concentration were legendary. He could recite poems, prayers and long passages from the Bible from memory and was himself an accomplished versifier. Towards the end of his life, when he was in charge of the Cavendish Laboratory a contemporary noted:

> When absorbed in his work he had a habit of whistling, not loudly, but in a half subdued manner, no particular tune discernible, but a sort of running accompaniment to his inward thoughts... and could pursue his studies under distractions which most students would find intolerable, such as loud conversations in the room where he was at work. On these occasions he used, in a manner, to take his dog into his confidence, and would say softly, 'Tobi, Tobi', at intervals...[63]

For a mathematical physicist, Maxwell was unusual in that he combined a powerful theoretical mind with great practicality. Throughout his life he was an inveterate and skilled experimenter and, never having lost the all-consuming curiosity of his childhood, was often engaged in seemingly trivial investigations such as whether cats invariably land on their feet or what causes a sheet of paper to flutter as it falls through the air.[64] A close friend noted that "...his mind could never bear to pass up any phenomenon without satisfying itself at least of its general nature and causes."[65] Maxwell brought with him to

---

[61] Campbell, L., Garnett, W. (1882), p 105.
[62] Campbell, L., Garnett, W. (1882), p 417.
[63] Campbell, L., Garnett, W. (1882), p 369.
[64] Campbell, L., Garnett, W. (1882), p 499.
[65] Tait, P. G. (1880), p 320.

Cambridge the experimental bric-a-brac that he had amassed in Edinburgh, "scraps of gelatins, gutta percha, and unannealed glass, his bits of magnetised steel, and other objects…".[66]

In keeping with the principle of mens sana in corpore sano,[67] which had become central to the ethos of Cambridge during the nineteenth century, Maxwell liked to keep fit walking and sculling. He was a keen if eccentric swimmer: he would throw himself face first into water in the belief that the impact stimulated his circulation. And for a short while he took to running along the corridors of his college in the early hours of the morning until his fellow undergraduates insisted that he stop.

During the nineteenth century, Cambridge's academic reputation rested on the Mathematical Tripos. All undergraduates had to sit and pass this lengthy mathematical examination in order to gain a degree, whatever subject they had studied. Strangely, university lectures did not specifically prepare students for the Tripos, so the more ambitious among them employed extracurricular tutors to coach them. The emphasis of instruction, however, was on passing the exam, not on producing original mathematicians. Those who gained first class degrees in the Tripos were known as Wranglers. Top Wranglers, i.e. the most promising mathematicians, went on to compete for the far more demanding Smith's prize. Maxwell was just beaten into second place in the Tripos by E.J. Routh (1831–1907), but was ranked joint first with him in the Smith's Prize, considered to be a far more reliable test of mathematical talent than the Tripos.[68]

Maxwell stayed on at Trinity and was elected a Fellow a couple of years after graduation. During this period he completed research on colour vision, a subject that had interested him since his Edinburgh days.[69] He also began work on the link between electricity and magnetism in which he had become interested as a result of reading Faraday's published work while still an undergraduate.

At that time, Maxwell was almost alone in taking Faraday's speculations on lines of force seriously. He realised that further progress was possible only if a mathematical theory could be developed to account for Faraday's experimental discoveries. The major problem was that the nature of these lines of force was unclear. Unable to tackle the problem directly, Maxwell had to come up with a physical analogy that was easier to understand, just as Young

---

[66] Campbell, L., Garnett, W. (1882), p 147.

[67] "A healthy mind in a healthy body".

[68] Routh was elected a fellow of Peterhouse, became a university tutor for the Tripos and later married one of Airy's daughters.

[69] Niven, W.D. (1890), Vol 2, pp 267–79.

had when he used the properties of sound waves to suggest those of light considered as a wave.

Maxwell's analogy seems at first sight ad hoc and distinctly unpromising. He pictured magnetism as a collection of spinning vortexes separated from one another by a layer of ball bearings, which represented electricity. Faraday's electromagnetic induction was represented by changes in the rate at which vortexes spin and which cause the ball bearings to move. Oersted's magnetic effect of an electric current was represented by a wholesale movement of the layers of ball bearings, which in turn affects the spin of the vortexes. In Maxwell's capable hands, however, this mechanical model yielded equations that supported Faraday's experimental results.[70]

But Maxwell wasn't interested in this mechanical model for its own sake. Despite the spinning vortices and rolling ball bearings and their interaction through direct contact, all so reminiscent of Descartes' explanation for light and colour, Maxwell was no latter-day Cartesian. The sole purpose of the analogy was to come up with equations that would predict results in line with Faraday's experimental discoveries. Once this was achieved, Maxwell discarded the model, enabling him to replace the network of Faraday's lines of force with the idea of a field of electromagnetic energy surrounding electrically charged bodies.[71]

But the most exciting feature of the model was that it predicted that any displacement in a layer of ball bearings creates a disturbance that travels through the imaginary arrangement of vortexes and bearings at the speed of light. Or, to use the vocabulary of electromagnetism, a rapidly fluctuating electric current causes a fluctuating magnetic field that in turn causes a fluctuating electric field that itself is the cause of a further magnetic field and so on ad infinitum (Fig. 9.3). The propagation of light can thus be described using equations that are similar to those that describe mechanical waves and so light can be considered to be an electromagnetic wave. What's more, the electromagnetic disturbance takes the form of a transverse wave and is therefore polarised, which as we saw in the previous chapter is a hallmark of light. Maxwell was fully aware of the implications:

> The velocity of transverse undulations in our hypothetical medium … agrees so exactly with the velocity of light calculated from the optical experiments of Fizeau that we can scarcely avoid the inference that light consists in the transverse undulations of the same medium which is the cause of electric and magnetic phenomena.[72]

---

[70] Mahon, B. (2003), pp 23–34.
[71] Mahon, B. (2003), pp 35–47.
[72] Niven, W.D. (1890), Vol 1, p 500.

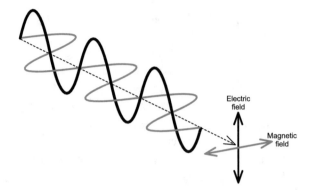

Fig. 9.3 Maxwell's electromagnet waves represented as sinusoidal waves

Surprisingly, given that Maxwell was a gifted experimentalist, he didn't take the obvious next step: experimental confirmation of his theory by creating an electromagnetic wave from an electric current. As we shall see, confirmation of his theory, almost a decade after his untimely death in 1879 at the age of 48, concerned not light but what we now call radio waves.

Maxwell didn't remain in Cambridge after gaining his Fellowship because in 1856 he was offered a post teaching physics in Aberdeen. He didn't find that city as congenial as Cambridge; soon after arriving there he wrote to a friend "No jokes of any kind are understood here. I have not made one for two months, and if I feel one coming I shall bite my tongue."[73] Maxwell proved to be a conscientious lecturer, but a poor teacher. The problem was that during his lectures he was easily side-tracked from the topic in hand and students found it difficult to follow his asides or keep up with the pace of his thinking. One of his contemporaries at Cambridge wrote "But while so ingenious himself, he had great difficulty in imparting his ideas to others; consequently was not so clear a lecturer or writer as might have been expected"[74]—echoes of Thomas Young. All the while in his spare time he continued to pursue his own researches and a year after arriving in Aberdeen he solved a problem that had defeated even that matchless master of celestial mechanics, Pierre-Simon Laplace: the nature of Saturn's rings. Maxwell's proof that they are composed of a vast number of small solid particles in orbit about Saturn has not been improved upon and has since been verified from images captured by the cameras of the Voyager 1 spacecraft in 1980. He also began work on the theory that explains the properties of gases assuming that they are composed of a large number of particles moving at random

---

[73] Campbell, L., Garnett, W. (1882), p 297.
[74] Campbell, L., Garnett, W. (1882), p 164.

with different velocities. His ideas in this field proved to be as scientifically important and significant as his work on electromagnetism.

A couple of years after Maxwell arrived in Aberdeen, he married Katherine Dewar (1824–86). She was the daughter of the college's principal, though this didn't help him when the college merged with another in 1860: Maxwell was made redundant. But later that year he was appointed Professor of Physics and Astronomy at King's College, London.[75] The five years he spent there were his most productive. He refined his theory of electromagnetism, laid down standards for electrical measurement, and made the world's first colour photograph. But his limitations as a lecturer continued to dog him and his reputation among his students at King's was no better than it had been in Aberdeen. The problem was that "the rapidity of his thinking, which he could not control, was such as to destroy, except for the highest class of students, the value of his lectures. His books and his written addresses (always gone over twice in MS) are models of clear and precise exposition; but his extempore lectures exhibited in a manner most aggravating to the listener the extraordinary fertility of his imagination."[76] He was also a poor disciplinarian, which added to the disorder of his lectures.

Maxwell was a countryman at heart, however, and the lure of Galloway eventually proved too much. He also wanted to be free to pursue his own work, so in 1865 he resigned his post at King's and went to live on his family estate with Katherine. But in 1871 he was persuaded to come out of retirement to oversee the creation of a research laboratory in Cambridge. The powers that be at the University had at last woken up to the fact that the march of science required more than clever mathematicians, it also demanded highly trained experimentalists. Under Maxwell, the laboratory—named the Cavendish Laboratory in honour of William Cavendish (1808–91), the 7th Duke of Devonshire and chancellor of the University, who put up the money—quickly established itself as one of the world's leading research centres, for although Maxwell was a poor lecturer, he was in his element in the one-to-one atmosphere of a laboratory, and was by all accounts an inspirational director of research.[77]

With Maxwell at the helm, might evidence for his electromagnetic theory of light been discovered at the Cavendish? It is an intriguing possibility, but one for which there is no support. His assistant at the Cavendish, William

---

[75] King's College London (KCL) was one of the two new universities set up in London. It was founded in 1829, three years after University College London (UCL).

[76] Tait, P.G. (1898), p 398.

[77] William Cavendish was an accomplished mathematician, having been second Wrangler in his year and awarded 1st Smith's Prizeman.

Garnett (1850–1932), wrote "I am confident that Maxwell never contemplated the experimental production of electromagnetic waves in the laboratory or he would have discussed ways and means with all his workers".[78] This was partly because the focus of Maxwell's interest in electromagnetism was almost exclusively on light. He acknowledged, in passing, that "radiant heat and other radiations, if any"[79] are also likely to be electromagnetic waves, but the note of uncertainty concerning "other radiations, if any" speaks volumes: he doesn't appear to have ever considered the possibility of creating electromagnetic waves by means of rapidly fluctuating electric currents.

Another reason why Maxwell failed to predict waves produced by purely electrical means may have been that he wasn't able to conceive of a method of creating light directly from electricity without the mediation of heat; all known sources of light at the time involved either a flame or a hot solid.[80] Moreover, in Maxwell's day, the known spectrum of radiant energy was limited to the visible spectrum, the near infrared and near ultraviolet, all of which are emitted by hot bodies.

The visible spectrum had been discovered by Newton in 1666—at least in the sense that he was the first to show that sunlight is not a single indivisible entity but is composed of parts, which, he found, differ from one another in terms of their refrangibility. But neither he, nor anyone else, suspected that there might be other forms of radiant energy until 1800, when William Herschel (1738–1822), the renowned Anglo-German astronomer who had discovered Uranus in 1781, noticed while observing the sun through a telescope fitted with different filters to protect his eyesight that "when I used some [filters], I felt a sensation of heat, though I had but little light; while others gave me much light, with scarce any sensation of heat."[81] Intrigued, he devised a simple experiment to investigate the heating effect of individual spectral colours: he placed the blackened bulb of a thermometer in different parts of the Sun's spectrum formed by passing a narrow beam of sunlight through a glass prism and found "…the heating power of prismatic colours, is very far from being equally divided, and … the red rays are chiefly eminent in that respect."[82] In a follow-up experiment he placed thermometers just beyond the edge of the visible spectrum and found that "…the maximum of the heating power is vested among the invisible rays" beyond the red end of the spectrum but was unable detect any heating effect beyond the violet end (Fig. 9.4). When he investigated the properties of these invisible warming

[78] Simpsom, T. K. (1966), p 432.

[79] Niven, W.D. (1890), Vol 1, p 535.

[80] Simpson, T. K. (1966), p 414.

[81] Herschel, W. (1800 a), p 256.

[82] Herschel, W. (1800 a), pp 261–2.

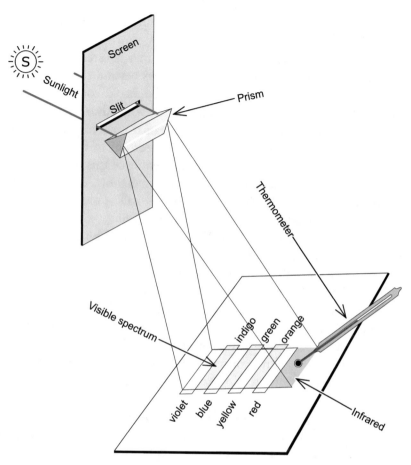

Fig. 9.4 Herschel's discovery of infrared radiation. Simplified copy of the experimental set he employed[84]

rays, he found that they appeared to have similar properties to visible light: they were emitted by hot bodies and could be reflected and refracted.[83]

Despite these similarities, Herschel eventually concluded that light and heat rays were different phenomena because "How can such effects that are so opposite be ascribed to the same cause? First of all, heat without light; next to this, decreasing heat, but increasing light; then again decreasing heat and decreasing light."[85]

[83] Lovell, D.J. (1968), p 50.
[84] Herschel, W. (1800 b), p 292.
[85] Herschel, W. (1800 c), p 508.

Herschel's discovery of invisible heat rays came to the notice of Johann Ritter (1776–1810), a leading *Naturphilosoph*,[86] shortly after its announcement. Being a *Naturphilosoph*, Ritter assumed that nature consist of opposites. So when he learned that there are heat rays beyond the red end of the visible spectrum he surmised that there must be a cold counterpart beyond its violet end. He knew that paper impregnated with silver chloride darkens more quickly when exposed to violet light than it does with other colours. Using a sheet of this paper as a screen for a spectrum of the sun's light, he discovered that the maximum amount of darkening of this paper occurs just beyond the violet end. Ritter named this new radiation "deoxidising rays". Later they were known as "actinic rays" and later still as ultraviolet rays.

It was that Cassandra of science, Thomas Young, who in 1801 first suggested that both Herschel's and Ritter's rays should be considered to be extensions of the visible spectrum.[87] Young claimed that heat rays should not be considered wholly distinct to light as Herschel believed because "it seems highly likely that light differs from heat only in the frequency of its radiations."[88] Herschel's heat rays are now known as infrared rays, i.e. they are so named because they lie beyond the red end of the visible spectrum. A year later Young described how he had used the "the Dark Rays of Ritter" to form a faint image of three concentric Newton's Rings on a sheet of paper impregnated with silver nitrate.[89] Here was evidence that Ritter's rays are also waves.

But the indifference of the scientific establishment to Young's wave theory at that time delayed for another decade a general acceptance of his suggestion that Herschel's and Ritter's rays are forms of light. And it was several decades before improvements in ways of detecting these invisible rays extended the radiant spectrum much beyond the fringes of the visible spectrum. The problem was twofold: mercury thermometers are not sensitive enough to detect infrared much beyond the red end the visible spectrum and the glass of a prism absorbs both infrared and ultraviolet strongly. In 1829, Leopoldo Nobili (1784–1835), an Italian natural philosopher known for his electrical researches, invented a far more sensitive detector of infrared known as a thermopile.[90] A few years later his colleague, Macedonio Melloni (1789–1854)

---

[86] Ritter introduced Oersted to *Naturphilosophie*.

[87] Cassandra was the daughter of Priam, king of Troy. She was given the gift of prophesy by Apollo, who had fallen in love with her. But she didn't return his love so he turned the gift into a curse by causing her prophesies to be disbelieved, even though they were true.

[88] Young, T. (1855), Vol I, p 168.

[89] Young, T. (1855), Vol I, pp 190–1.

[90] A thermopile converts heat into electricity and the resulting current can be detected with a sensitive galvanometer.

Fig. 9.5 Multispectral rainbow. These three images are of the same bow using i.r. and u.v. filters. The central portion is of the visible light bow, below this is the infrared bow and above it is the ultraviolet bow[91]

discovered that rock salt is far more transparent to infrared than glass—hence the use of prisms made from rock salt in modern spectroscopy. And in 1862, Gabriel Stokes (1819–1903), then Lucasian Professor, discovered that quartz is transparent to ultraviolet, which enabled him to detect this radiation far beyond the violet edge of the visible spectrum (Fig. 9.5).

Given Maxwell's apparent lack of interest in an electromagnetic test of his theory, Cambridge was the last place where evidence for his electromagnetic theory was likely to be discovered even if he had not died prematurely in 1879 from an abdominal cancer, the same cancer that had killed his mother forty years earlier. In any case, at the time of his death, there was little interest in his ideas outside a small group of Cambridge scientists. So with Maxwell gone there was no one of sufficient scientific stature in Britain to promote his ideas or make research students aware of the implications of his electromagnetic theory. And so the Cavendish missed the opportunity to make one of the great scientific discoveries of the nineteenth century: radio waves.

Maxwell's ideas came to the attention of Herman von Helmholtz (1821–1894), one of the movers and shakers of the generation of German scientists that had managed to free itself from the wilder shores of *Naturphilosophie*

---

[91] Photo by Andrew Steele, used with permission.

and rejoin the Newtonian mainstream. As a youth Helmholtz had wanted to study physics but in order to fund his studies he had had to train as a doctor in order to have his fees paid by the Prussian state. In return he was required to serve in the army for several years following graduation. His first academic post was as a physiologist and his earliest scientific work was on the conservation of energy, which, given his medical background, he approached in terms of work done by the muscles of that unwitting martyr to science, the frog. An interest in physiology led him to study vision and hearing, which included coming up with an account of colour vision along the lines first suggested by Thomas Young.[92] Indeed, Helmholtz and Young had much in common: both were polymaths, had trained as physicians and made ground-breaking discoveries in both physiology and physics.[93]

One of Helmholtz's protégés was Heinrich Hertz (1857–1894). As a boy, Hertz had his own workshop complete with a workbench, hand tools and a lathe, with which he made a number of optical instruments, including that all-time favourite of every tyro scientist, a telescope. He was also a gifted linguist, both of ancient and modern languages: he came top in Greek at his school and took private lessons in Arabic. He seemed destined for a career in engineering, but a growing interest in mathematics and physics led him to Berlin University and Helmholtz. Having obtained his doctorate, he embarked on an academic career and did the rounds of several of Germany's leading universities. By all accounts, he was an inspiring teacher, but eventually the engineer in him could no longer be denied and he abandoned the mathematical abstractions of theoretical physics for the tangible pleasures of the laboratory.

Hertz struck gold in 1888 when, quite by chance, he succeeded in producing radio waves in a laboratory. He had been set on the trail by Helmholtz, who had been among the first physicists to realise that the ideas of Faraday and Maxwell gave a far better account of electromagnetic phenomena than the action-at-a-distance school that prevailed in Germany at that time. In 1879 he persuaded Berlin University to announce a prize for an experimental proof of Maxwell's electromagnetic theory. But Hertz had not been searching for electromagnetic waves. Maxwell had predicted that one of the consequences of his electromagnetic field should be the presence of brief electric currents in the space between charged bodies. Hertz was attempting to detect these currents by subjecting a slab of wax to a high frequency electric current when he realised that the unwanted and troublesome sparks

---

[92] Helmholtz, H. (1962), pp 141–5.

[93] Helmholtz's interest in the perception of sound and music resulted in the most influential works ever published on the subject. See: Helmholtz, H. (1895).

that occurred in other parts of his apparatus, and which he found all but impossible to prevent, indicated the presence of one of Maxwell's "other radiations".

The unwanted sparks were due to currents induced in the metallic parts of the apparatus by electromagnetic waves created by the primary spark of the high frequency current. Hertz realised that here was a reliable method for detecting an electromagnetic wave: its presence would be signalled by a spark in the gap of a coil of wire several metres away. He abandoned the original experiment and set out in search of Maxwell's hitherto unsuspected electromagnetic wave.

If Hertz was to prove that his apparatus was indeed creating an electro-magnetic wave, he had to demonstrate that the wave had similar properties to light. From measurements of its frequency and wavelength he was able to calculate the wave's speed: it proved to be the same as that of light.[94] Further experiments established that the waves were polarised and could be reflected, refracted and diffracted, just like light.[95] Maxwell's theory had been vindicated and the spectrum of radiant energy had been extended well beyond infrared for Hertz's radio waves had a wavelength that was millions of times greater than that of light. Oliver Lodge (1851–1940), a British physi-cist working in Manchester, and who was on the verge of making the same discovery as Hertz, recognised that physics had arrived at a watershed: "The whole domain of Optics is now annexed to Electricity, which has thus become an imperial science."[96]

The effects of low frequency radio waves had, in fact, been noticed on several occasions before 1888 because even the slightest electrical spark will induce a current in a conductor some distance away. Indeed, Galvani's elec-trical research had originally been stimulated by a chance observation of this effect during the dissection of a frog's leg by one of his students: elsewhere in the laboratory, another student was operating a generator of static elec-tricity and noticed that every time the generator produced a spark the frog's leg contracted involuntarily.[97] In the light of Hertz's discovery, we now know that the generator's spark created short-lived, low frequency radio waves that induced a tiny current in the fluid of the frog's leg, causing the muscle to contract. Several weeks after this observation, Galvani noticed a frog's leg

---

[94] The frequency of the wave was that of the spark, 45 MHz. Its wavelength was determined by reflecting the wave to create a standing wave, and was 6.6 m. Frequency multiplied by wavelength equals velocity: $45{,}000{,}000 \times 6.6 = 300{,}000{,}000$ m/s (which is the speed of light).

[95] Hertz, H. (1888).

[96] Lodge, O. (1907), p 289.

[97] Galvani, L. (1791). English translation in: Magie, W.M. (1963), pp 421–7.

twitching in response to a stroke of lightning, just as it had in his laboratory. But, of course, he had no way of knowing that a frog's leg was acting as a detector of radio waves created by the rapid surge of electric charge that is the cause of a lightning stoke.

Galvani spent several fruitless months trying to make sense of these observations before turning his attention to the phenomenon he is known for: the electric current that flows in a frog's leg when it is in contact with two dissimilar metals. The phenomenon of induced currents was later studied in great detail in 1842 by Joseph Henry (1797–1878). Henry was acclaimed during his lifetime as one of America's foremost experimentalists, and with good reason: much of his research into electricity and magnetism paralleled Faraday's and indeed equalled it in some respects. Henry had even speculated that the current induced in a conductor was the result of an electric disturbance created by an electric spark and which travelled through the intervening space. But isolated from the centres of European science by the Atlantic Ocean, Henry's ideas failed to strike a chord among the leading European scientists of his day. Nevertheless, Henry's discoveries were known in Britain, and had Maxwell sought a means to prove his theory through experiment, he could have made use of Henry's research.[98]

In 1895 Maxwell's electromagnetic spectrum acquired another member. Wilhelm Röntgen (1845–1923), a German physicist, was investigating the effects of a beam of electrons travelling through a vacuum within a glass vessel when he noticed that a nearby screen coated with a fluorescent material glowed when the apparatus was in action. Further investigation convinced him that he had discovered a new type of radiant energy, which he called X-rays—though everyone else at the time preferred to call them *Röntgen-rays*. The most remarkable property of these rays is that they pass through opaque matter, and a month after his discovery Röntgen used a beam of X-rays to make an image of his wife's hand that showed the bones as a dark silhouette within the faint outline of the hand's flesh. The fact that he had used an ordinary photographic plate to capture the image was taken as evidence that a new form of light had been discovered, but the rays showed no sign that they shared light's other properties.[99] For several years after their discovery no one succeeded in reflecting or refracting them and it wasn't until 1912 that anyone succeeded in showing that they diffract and cause interference when they pass through the regular arrangement of atoms within a crystal.[100] Thus for a long while after their discovery X-rays were considered to be a series of

---

[98] Simpson, T.K. (1966), p 417.
[99] Röentgen, W.K. (1895), pp 132–141. English translation in: Magie, W.M. (1963), pp 600–10.
[100] Keller, A. (1983), p 182.

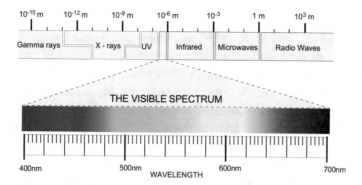

Fig. 9.6 The electromagnetic spectrum. The wavelength of visible light is approximately one half millionth of a millimetre

solitary electromagnetic pulses reminiscent of Huygens ideas on the nature of light, rather than a periodic wave.

In 1900, five years after Röntgen's discovery, yet another form of electromagnetic radiation, even more penetrating than X-rays, was discovered by Paul Villard (1860–1934), a French chemist, while he was studying the radiation emitted by radium. Villard's radiation was named *gamma radiation* by Ernest Rutherford (1871–1937), who had earlier established that radium emits two types of particle, and which he had named Alpha and Beta particles (Fig. 9.6).

Although Hertz's discovery vindicated Maxwell's conjecture that light and 'other radiations' are electromagnetic waves, there was another aspect of Maxwell's theory that had yet to be properly explored. This was the nature of the æther, which Maxwell and everyone else at the time believed was the medium necessary for the propagation of these waves. It proved to be a chimera. In the first place, beginning in 1881, in a series of increasingly more precise experiments based on a suggestion by Maxwell, A. A. Michelson (1852–1931), an American physicist of Polish extraction, failed to detect its existence.[101] The experiment involved comparing the speed of two beams of light from the same source, one parallel the direction in which the Earth orbits the sun and the other perpendicular to this. The idea was that any difference between the speeds would be due to the earth's motion through the æther. None of Michelson's experiments detected the slightest difference, though they did result in a very precise determination for the speed of light.

In 1905 a far more fundamental attack on the idea of the æther was launched by Albert Einstein (1879–1955), then an obscure clerk working for

---

[101] Niven, W.D. (1890), Vol 2, pp 763–75.

the Swiss Patent Office in Berne. Einstein was born in Germany but had gone to Switzerland in 1895 to complete his education. A year later he renounced his German citizenship in order to avoid conscription. He graduated in physics from Zurich Polytechnic in 1900 with plans to become a school teacher. But he had made a poor impression on his professors and did not perform particularly well in his finals. So while everyone else in his year found academic positions, he was left on the shelf. His extreme independence of mind and character, so important to his later success as a scientific iconoclast, made him a poor student, at least in the eyes of his teachers. He admitted that he was as stubborn as a mule; he was also arrogant and delighted in vexing his professors. He skipped classes and made no secret of his contempt for lecturers he believed were out of touch with the latest ideas. One of his lecturers, the eminent mathematician Hermann Minkowski (1864–1909), dismissed Einstein as a "lazy dog"—though in Einstein's defence it should be said that Minkowski was a notoriously uninspiring teacher and made no allowance for difficulties experienced by his students.[102] Moreover, Einstein had no great fondness of mathematics. Rather like Newton at Cambridge, Einstein largely ignored the prescribed syllabus and made a private study of physicists that interested him, particularly of Maxwell. Having fallen foul of his professors, he was unable to find anyone to give him a decent job reference. He was fortunate to get work with the patent office a year after graduating.

The isolation from academic circles that life as a patent clerk entailed was a blessing in disguise for it gave him the freedom to pursue his own ideas; and he had several of those. Finally, in 1905, he was ready to share them with the scientific world. That was the year in which he published three of the most influential papers in the history of physics. The first of these provided incontrovertible proof for the existence of atoms, which even at that late date were considered by the majority of physicists to be, at best, no more than theoretically useful entities. Another laid the foundations of his Special Theory of Relativity. But the most revolutionary of his publications that year was the paper that dealt with the nature of light in which he argued that in some circumstances light behaves like a particle: it was to shake physics to its foundations and lead directly to the creation of quantum physics.[103]

Einstein arrived at his theory of relativity from a study of Maxwell's theory of electromagnetism. In Maxwell's theory, the speed of light has a fixed value that is the same for all observers in all situations. Moreover, this is neither an ad hoc assumption on Maxwell's part nor is it based on a direct measurement

[102] Minkowski had been taught physics by Heinrich Hertz.
[103] Stachel, J. (1998), for English translations of these papers.

of the speed of light. Maxwell had derived the speed of his electromagnetic waves from measurements of electrical and magnetic constants; as he pointed out somewhat facetiously when discussing the experimental basis of the theoretical value he had calculated for the speed of electromagnetic waves: "The only use made of light in the experiment was to see the instruments."[104] But two hundred and fifty years earlier Galileo had proved that it is only possible to measure speed if this is done relative to a particular frame of reference, an idea that was later incorporated into Newtonian physics. Maxwell had assumed that where light is concerned, that frame of reference is the æther. In Newtonian physics, however, the speed of light has no special status and therefore depends on the relative speed between the source and observer—that was the assumption behind Michelson's experiment.

Einstein agreed with Maxwell that the speed of light is independent of all frames of reference. At the same time he accepted Galileo's principle of the relativity of speed. Working through the consequences of reconciling these two apparently contradictory assumptions, he turned the Newtonian world inside out and showed that measurements of time and distance vary depending on the relative speed between the observer and the observed. One of the most paradoxical, not to say perplexing consequences of the relativity of space and time is that it is not possible for widely separated observers to agree on the order in which distant events occur, which, of course, is completely contrary to our everyday notion of cause and effect. Fortunately, this Alice-in-Wonderland world is evident only when the relative speed of observer and observed is close to the speed of light, so we are fortunately spared the disorientating effects of Einstein's relativity in daily life.[105] As for Maxwell's æther, by taking the speed of light as a universal constant, Einstein declared that it is "…superfluous, inasmuch as the [Special Theory of Relativity] will not require a 'space at absolute rest' endowed with special properties…".[106]

Einstein's Special Theory of Relativity was the capstone of Maxwell's electromagnetic theory and so, for all its novelty, which meant that it was some years before it was widely accepted, it doesn't represent a radical break with classical physics, as his do Einstein's ideas about the nature of light.

Those ideas had their origin in a curious phenomenon that Hertz had noticed in the course of his electromagnetic experiments, one that proved to be far more significant to developments in physics during the twentieth century than his discovery of radio waves. The spark induced in his detector was not very bright and in order to see it more easily he sometimes shielded

---

[104] Niven, W.D. (1890), Vol 1, p 580.

[105] Gamow, G. (1993), for an amusing and instructive insight into the effects of relativity.

[106] Stachel, J. (1998), p 124.

it from the laboratory lights by enclosing it in case. When he did so he found that the spark was even weaker. On further investigation he found that the spark in the detector was enhanced when the detector was directly illuminated by ultraviolet light produced by primary spark. But having noted the phenomenon he didn't follow it up. A couple of decades later, this apparently trivial phenomenon was to lead to the most fundamental revolution in physics since the publication of Newton's Principia.

By the close of the nineteenth century, when every scientist had come to accept that light is a wave, and when it seemed as if all that remained in physics was to dot the i's and cross the t's, another German physicist, Max Planck (1858–1947), inadvertently stumbled across something nasty in the woodshed. In 1900, in an attempt to explain the fact that the observed distribution of the different forms of electromagnetic radiation emitted from a hot surface was at odds with the accepted physics of the day, the so-called *black-body problem*, Planck suggested that it would be a useful stopgap to assume that electromagnetic energy is absorbed and emitted in discrete bundles, to which he gave the name quantum of action, rather than as continuous waves. His hunch proved right, though Planck was sure that the assumption that light is quantized would eventually prove unnecessary when the phenomenon of *black-body radiation* was better understood.

Five years later, in 1905, Einstein employed Planck's quantum of action to explain the phenomenon that had first been noticed by Hertz: the enhanced discharge of electrons from an electrically charged body when it is illuminated with ultraviolet light, the so-called photoelectric effect. In the meantime, further investigation of the phenomenon by others had discovered another puzzle: the emission of electrons by light depends only on the frequency of light and not on its intensity.[107] Furthermore, there always is a threshold frequency, which varies from one substance to another, below which no electrons are emitted. So, for example, no discharge of electrons occurs when a particular substance is illuminated by a bright red light but some are emitted when illuminated by a dim blue light—in the electromagnetic theory the energy of light is determined by its frequency and the frequency of blue light is approximately twice that of red light. These observations strongly suggested that all the energy of the incident radiation is either absorbed instantaneously or not at all. But if light is a wave then one would expect that, given enough time, electrons would be emitted whatever the frequency of the illumination because an electron would eventually accumulate enough energy from the incident wave for it to break free from its parent atom. However, as the

---

[107] Wheaton, B. R. (1978), p 299.

experiment with red light demonstrates, this never happens. Einstein showed that all these facts could be explained by assuming that in some circumstances light really is quantised, i.e. that in the case of the photoelectric effect it acts as if it is a particle rather than a wave. His explanation for the photoelectric effect was experimentally confirmed in 1915 by a sceptical American physicist: Robert Millikan (1868–1953). Millikan set out to refute Einstein only to discover that he was correct. Einstein was awarded the Nobel Prize for this work in 1922. And four years later, in 1926, an American chemist, proposed that Einstein's quantum of action be renamed "photon".[108]

Further evidence for Einstein's explanation of the photoelectric effect in terms of photons was found in 1923 by another American physicist, Arthur Compton (1892–1962). By that date, the discovery that X-rays can be diffracted showed that they too are waves. But Compton found that when an X-ray collides with an electron it loses energy in exactly the same way as it would if it were a particle. In the same year, the idea that waves can behave like particles prompted a young Frenchman, Louis de Broglie to ask himself: if light, which we know to be a wave, also behaves as if it is a particle, why shouldn't an electron, or indeed any material particle, also behave like a wave? He soon came up with an answer.

Louis de Broglie (1892–1987) was the direct descendant of Victor-François de Broglie, on whose estate Augustin Fresnel was born.[109] As a young man, de Broglie was more interested in history than science and the expectation was that he would become a diplomat but, influenced by his elder brother, Maurice, he eventually took up physics. Maurice was a distinguished physicist and had a well-equipped private laboratory in his home in which he worked on the properties of X-rays and the photoelectric effect. Undoubtedly it was Maurice's photoelectric experiments that set Louis thinking. A year after posing his question he submitted his doctoral thesis to the Sorbonne in which he made use of Einstein's mathematical theory of photons to propose that in some situations very small particles of matter such as electrons should behave like waves. One of the examiners, who initially considered the idea far-fetched, consulted Einstein before passing judgement. Einstein replied: "He has lifted a corner of a great veil".[110] Reassured, though none the wiser, the examiners awarded de Broglie his doctorate. Someone who was present at the public examination of de Broglie's thesis recalled "Never had so much gone over the heads of so many!".[111]

---

[108] Lewis, G. (1926).

[109] Victor-François de Broglie (1718–1804) was Louis de Broglie's great, great, great, great grandfather.

[110] Abragam, A. (1988), p 30.

[111] Abragam, A. (1988), p 30.

As for an experimental test of his hypothesis, de Broglie suggested that the passage of electrons through a crystal lattice should produce a diffraction pattern similar to that created by X-rays, the very effect that Fresnel had claimed is the hallmark of light waves. In fact, the earliest confirmation of de Broglie's hypothesis occurred in 1927 as the result of a laboratory accident. Unaware of de Broglie's ideas, an American physicist, Clinton Davidson (1881–1958), had been investigating the effect of firing electrons at a metal target within a vacuum and had had to repair his apparatus after it was damaged when the glass enclosure broke. This involved heating the target to clean its surface. Unknown to Davidson, this created a uniform crystalline lattice on the target's surface which, when he restarted his investigation, caused the reflected beam of electrons to produce a diffraction pattern. He duly published his results and, unaware of the importance of his discovery, went on to other things. It was only when he visited Britain the following year to attend a scientific conference that someone pointed out to him that his results were in line with de Broglie's predictions. Similar results were later obtained by George Thomson (1892–1975), the son of J. J. Thomson (1856–1940), the Cavendish Professor who had discovered the electron in 1897. Both men were awarded Nobel Prizes: the father in 1906 for proving that the electron is a particle, when it was widely believed to be a form of radiation, and the son in 1937 for proving that it is a wave!

The era of wave-particle duality had arrived and found the scientific establishment unprepared. Nothing in our everyday experience enables us picture the electron as a particle-cum-wave or a photon as a wave-cum-particle. Physicists are saved from complete bewilderment by the mathematical equations of quantum theory that describe the properties of these entities. But the rest of us may well feel cheated, unreasonably as it happens, for when we set out to get to the bottom of even the most mundane object or event we are confronted with what the philosopher Iris Murdoch (1919–99) identified as the "inexhaustible detail of the world",[112] a situation that can often strain one's imagination to breaking point.

We now accept that it is no longer possible to say unambiguously what light is, only that it acts like a particle (i.e. photon) when it is emitted or absorbed by matter and that it travels through space as if it were a wave. It would be a mistake, however, to imagine that Newton has been vindicated because photons are not Newtonian corpuscles. In order to explain phenomena such as refraction, interference, diffraction and polarisation, Newton assumed that his corpuscles have mass and exert forces on matter. A

---

[112] Murdoch, I. (1998), p 86. Iris Murdoch is referring to human relationships, not the physical world, but what she says applies equally to the material world.

photon, however, far from being an entity is actually a process.[113] It describes how light interacts with matter whereas refraction, interference, diffraction and polarisation are effects that occur only when light is travelling through space, and so are due to its wave properties.

Descartes was right: there really is no direct correspondence between the world as we perceive it and the reality that underpins it. But quantum theory widens the Cartesian chasm between experience and reality. Not only is the world devoid of colour, warmth, sound and odour, one the certainties of the mechanical philosophy—that the world is made of matter, of some-thing tangible—is itself consigned to subjective experience. We are left on the outside with our noses pressed against the window pane, forever excluded from what lies within.

And what of the rainbow in the quantum era? In 1908, Gustave Mie (1869–1957) published a comprehensive theory of the interaction between light and matter based on Maxwell's electromagnetic theory. Although Mie did not apply his theory to the rainbow, others have done so.[114] The result is a far more accurate picture of the distribution of light and colour in the rainbow that throws up details invisible to the human eye. Theory had at last caught up and overtaken the rainbow, which, unsurprisingly, now ceased to be a useful touchstone for theories of light or to engage leading scientists. Moreover, the electrification of the rainbow has not transformed our under-standing of the phenomenon as much as Young's and Fresnel's mechanical light wave did.

Surprisingly, following Hertz's discovery of radio waves it seems that it took another 90 years before anyone seriously considered the possibility that if light is an electromagnetic wave and forms rainbows, then it is reasonable to expect that other types of electromagnetic radiation can do the same: infrared and ultraviolet ones, if not radio waves or X rays. Of course, we cannot see these any of radiations, we require instruments to detect their presence. In 1971, an American physicist, Robert Greenler (b. 1929), took the first photo-graph of an infrared bow using infrared sensitive photographic film. The arc of this bow lies just beyond that of the visible red arc, and like a visible light bow, it is accompanied by supernumerary arcs. Since Greenler's pioneering efforts, the ultraviolet bow has also been imaged (see Fig. 9.5).

---

[113] Interactions between light and matter occurs through electrons, i.e. charged particles.
[114] Lee, R.L. and Fraser, A.B. (2001), pp 314–16.

Even though mainstream science no longer has much interest in the rainbow, the phenomenon continues to fascinate. Has the third rainbow ever been seen in nature? Or, indeed, rainbows of higher orders? How to explain reports of rainbows with variable curvature? Or those that appear to split in two? Are natural rainbows seen only in drops of water? We'll address these and other questions in the next chapter.

10

In the Eye of the Beholder

We see nothing truly till we understand it[1]

Physics may long ago have predicted that there is much more to the rainbow than the primary and secondary ones we are familiar with, that the curvature of the bow is not necessarily circular (because raindrops larger than a certain size are not spherical), and that rainbows of different curvature can occur at the same time. But do such bows actually exist? Have they been seen in nature?

Add to this that before the advent of photographic film capable of taking colour photographs, there were any number anecdotal accounts of rainbows that physics seemed unable to explain. The misplaced confidence that one must necessarily see what is before one's eyes, together with an ignorance of the nature of what is seen led to innumerable reports of impossible rainbows, to which one may add confusing them with ice halos in their various forms.[2] A photograph of an unusual rainbow—indeed, of any rainbow—is worth a thousand words, and can be studied, measured and compared with what is already known to establish its authenticity.

Consider the case of the tertiary rainbow. According to Aristotle "Three or more bows are never seen, because even the second is dimmer than the first, and so the third reflection is altogether too feeble to reach the sun."[3]

---

[1] Leslie, C.R. (1896), p. 393.

[2] Corliss, W.R. (1983), pp. 111–25.

[3] Aristotle (1952), p. 267.

Yet, far from being the last word on the subject, the mere mention of three or more bows led to centuries of debate concerning their existence. Sometimes denied, at others accepted as possible, the problem was always that without a proper understanding of the optics of a rainbow, no one knew where to look for it. Given Aristotle's explanation for the primary and secondary bow, it was widely assumed that a third bow, were it possible, would likely be seen above the secondary bow, and therefore on the opposite side of the sky from the sun.

It's correct position in the sky couldn't be established until the path of light through a drop was precisely determined following the discovery of the law of refraction by Descartes in the seventeenth century. We saw in chapter six that both Newton and Halley later used that law to calculate the path of light through a rainbow that would give rise to a third rainbow and, indeed, to a fourth one as well. But neither man ever got around to looking for these bows, believing that because they are formed around the sun, the brightness of the sky in that direction would render them too faint to be visible.

Nevertheless, since then, there has been the occasional chance sighting of this elusive bow.[4] And as if to confirm just how elusive it is in nature, the first photograph of a tertiary bow was obtained in 2011 by pointing a camera in the right direction and taking several photographs without the bow itself being visible. Subsequent digital processing of the images revealed the bow. We now have unequivocal photographic evidence that the tertiary rainbow can occur in the open air. Both the quaternary (formed between observer and sun) and quinary rainbows (formed between the primary and secondary bows, i.e. within Alexander's dark band) have also been captured on film, though, as with the tertiary bow, heavy digital processing of the images was necessary to reveal them.[5]

These faint bows have long been studied in the laboratory, where it is possible to completely exclude the sun's glare and the sky's brightness. The first person to do this was Jaques Babinet (1794–1872), a French physicist, mathematician and astronomer. In 1837 he illuminated a narrow, vertical stream of water with sunlight, moonlight and candle light. "Pour water, alcohol, sulfuric acid, ether, salt water, through a cylindrical hole made in a piece of glass or metal, you will reproduce the meteorological angles known for water, and moreover you will see the third and the fourth rainbow at least, if the diameter of the liquid cylinder is not too small."[6] Of course, Babinet did not see rainbow arcs, he saw bright, multicoloured spots of light, as had

---

[4] Naylor, J. (2002), pp. 122–3.

[5] Haußmann, A. (2016), pp. 20–1.

[6] Babinet, J. (1837), pp. 645–8.

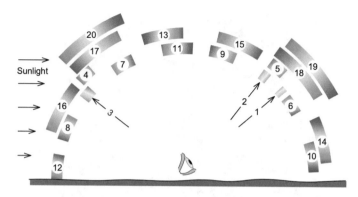

Fig. 10.1 A modern take on Billet's Rose. 1 is the primary rainbow, 2 the secondary, 3 the tertiary, 4 the quaternary etc. Note that the arc of successive rainbows grow wider and fainter and that the spectrum is reversed from one bow to the next[8]

Descartes when he held up that bowl of water above the shadow of his head on a sunny day some two hundred years earlier.

A more systematic study of higher order rainbows was published in 1868 by Felix Billet (1808–1882), a French physicist. Borrowing Babinet's method of illuminating a thin vertical stream of water and using the sun as the source of light, Billet was able to see 19 distinct rainbow orders, including their supernumerary arcs, albeit as rectangular, multicoloured bands, not as curved arcs. He presented the results in a figure as a series of spectra of different widths arranged around the horizontal, circular cross section of the stream of water. He named it *la rose des arc-en-ciel*, known in English as *Billet's Rose* (Fig. 10.1).[7]

Tertiary rainbows are as rare as hen's teeth, so they are very unlikely to catch the eye of the casual observer. They are not a source of reports of rainbows that appear to defy explanation. That distinction goes to rainbows formed by reflected sunlight.

Although the earth is illuminated by a single sun, in the right circumstances rain can be illuminated by two. One of these is the sun itself, the other is its reflection. Light from the reflected sun reaches the rain from a point below the horizon and forms a bow that encircles the primary bow and meets it at the horizon (Fig. 10.2).

The angle between the direct bow and reflected light bow increases until the sun is 42° above the horizon, at which point no primary bow is visible from the ground and, in theory, the reflection bow should be a full circle

---

[7] Billet, F. (1868), pp. 67–109.

[8] Walker, J., (1977).

Fig. 10.2 Reflected light bows over River Foyle, Derry, N.I. Apart from the secondary reflected light bow all the others are themselves reflected in the river which makes 7 bows in all![9]

between the ground and the zenith. There are, however, no records of circular bows due to reflected sunlight because the brightness of this rainbow decreases as the sun rises in the sky. The reason is that the amount of light reflected by water is greatest when the sun is at the horizon and diminishes rapidly as it rises. Well before the sun rises 42° above the horizon its reflection bright enough to create a rainbow that is visible. At the same time, the surface of the body water must be smooth enough to ensure that there are no multiple reflections of the sun, which would cause the reflection bow to look blurred.

Something else that affects the brightness of the reflected light bow is that light is partially polarised when reflected by water. The plane of this polarised light is parallel to the reflecting surface. We saw in chapter eight that the arc of a rainbow is tangentially polarised. Hence light polarised in a horizontal plane can't contribute to the brightness of the reflected bow where it meets the ground. A reflection bow is thus brightest at its apex and faintest at its foot.

---

[9] Photo by William Bradley, 3rd November, 2020, used with permission.

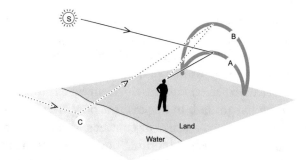

Fig. 10.3 The geometry of a reflected light rainbow. A is the primary arc due to direct sunlight and B is the primary arc due to sunlight reflected at C from a body of water between the observer and the sun

Halley, as we noted in chapter two, was the first person to publish an account of a reflected rainbow. One of the aspects of this extraordinary sight that caught his attention was that "the Secondary Iris lost its Colours, and appeared like a White Arch at the Top."[10] The explanation is that the sequence of colours in a reflected bow is the same as that of a primary bow and therefore the inverse of those in a secondary bow, so where the colours of the reflection bow coincide with those of the secondary bow, only a white patch is visible.

Reflected sunlight rainbows have given rise to more reports of strange and impossible rainbows than any other aspect of the rainbow, not least because in most situations when they are seen, it is often only the lower section of the arc that is visible. Hence there are numerous reports of bifurcated, V-shaped rainbows made by people unaware that what they have seen is two primary bows rather than a single abnormal one (Fig. 10.3).

Reflection bows are not the only example of bifurcated rainbows. In recent times photographs have established the existence of what have come to be known as twinned bows. The hallmark of a twinned bow is that the primary bow splits in two near the apex of the arc so that two closely spaced primary bows become visible, if only briefly. Twinned bows are, as you might expect, rather rare and are often taken to be reflection rainbows (and vice versa).

As for their cause, the explanation was available long before they were discovered. As we saw in chapter seven, Giambatista Venturi had investigated the effect on the curvature of a rainbow of drops that are not spherical and concluded that the presence of oblate drops in a shower of rain made up of spherical drops should create a second arc just below the apex of the primary bow. Hence one would expect to see a twinned bow in a rain shower

---

[10] Halley, E. (1698), p. 194.

composed of both spherical drops and oblate drops. Simulations of such a scenario suggests that this may well be the correct explanation.[11] Twinned bows are frequently confused with reflected light bows, but the difference between them is usually obvious in a photograph.

Perhaps the most unexpected circumstances in which to see a rainbow is not in the sky but on the ground. In this case the bow itself is formed in drops of dew deposited on either grass stalks or cobwebs. If the area on which dew is deposited is sufficiently large, the resulting bow forms a truncated oval with its apex near the observer's feet. Dewbows are also known as ground bows and are much fainter than bows seen in the sky because the layer of drops in which they are formed is extremely shallow, so its colours are barely discernible.

Like all rainbows, it too is due to light that enters your eye from a multitude of drops. Unlike the three-dimensional distribution of raindrops necessary for a rainbow, however, these drops all lie on a horizontal surface. Sidestepping the complex geometry necessary to explain the precise shape of such a bow, imagine what a horizontal slice through a primary rainbow would look like. Since a rainbow is a cone of light with your eye at its apex, a horizontal slice through this cone will be a hyperbola with its apex close to your feet. As you might expect, a ground bow will be extremely faint compared to the common or garden rainbow because it is formed in a very thin layer of drops, so you really have to have your wits about you if you are to notice it.

You are most likely to see a ground rainbow early in the morning during autumn as a fog clears. A grass lawn covered in fine spider's web, or gossamer, makes an ideal surface. Drops from the fog are deposited on the fibres of gossamer. As the fog clears, and the drops are illuminated by a low Sun, a faint arc of a ground rainbow may be visible. Although these bows are often called dewbows, this is a misnomer since the origin of the drops is fog rather than dew. Here is a delightful account of such a bow seen by a golfer (Fig. 10.4).

> The place, the conditions and the time were as follows. On Friday, November 5th, at Tadmorton Heath golf course, situated on the high ground near Banbury, there was a thick fog which persisted until about 11.30 a.m. in spite of a Sun which seemed to be struggling to penetrate it. The ground, more particularly noticeable on the fairways, was literally covered with a dense carpet of cobwebs which, being saturated either with particles of moisture or the remains of a ground frost - I do not know which - presented an appearance suggestive of snow. At 11.45, when the fog had suddenly dispersed and a brilliant Sun was almost due south, my companion and I came to a fairway running directly north and, upon stepping on to the mown and heavily

---

[11] Cowley, L., twinned bows, https://tinyurl.com/mr2hr6nd. Accessed 06/09/22.

Fig. 10.4 Ground rainbow. The brightness around the shadow of the photographer's head is known as a heiligenschein and is due to light reflected by drops of water resting on blades of grass[12]

cobwebbed portion, were simultaneously arrested by the extraordinary spectacle of a perfect rainbow flat on the grass. It started at our feet and ran in an elliptical form, definitely not a circle, to the extremities of both sides of the fairway and continued in front of us until the lightly mown turf of the green was reached. It was faint compared with the brilliance of an ordinary rainbow in the sky, but in all other respects was identical, and, culminating as it did at our feet, it added appreciably to the recognised difficulties of hitting a golf ball. In all it lasted for, or was discernible when our backs were to the sun, about an hour and a half.[13]

Colourless, horizontal bows are occasionally seen when looking down on an extensive layer of cloud, say from an aeroplane flying above it. Cloudbows, as they are known, are in fact horizontal fogbows, and so, like ground bows, the apex of the arc is closer to the eye than its ends. However, a cloudbow is only marginally brighter than the clouds in which it is seen, but since its light is polarised, looking through a polarising filter and rotating it will make the bow stand out from the cloud layer in which it forms.[14]

We associate rainbows with sunlight and rain, but all that is necessary to see a rainbow is a source of bright light such as a searchlight or even a powerful torch and a curtain of small, transparent drops of liquid. Searchlight rainbows are seen within the narrow beam of the light as short multicoloured arcs.[15]

---

[12] Photo by Lesmalvern, Wikimedia Commons. https://tinyurl.com/4cxn282u. Accessed 19/10/22.
[13] Wills, G.H.A. (1937).
[14] Können, G. P. (1985), p. 53.
[15] Cowley, L., searchlight rainbows, https://tinyurl.com/2873v62j. Accessed 06/09/22.

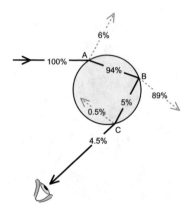

Fig. 10.5 Rainbows are nowhere near as bright as we assume they are. Slightly less than 5% of the light responsible for the rainbow ray that enters a drop at A reaches the eye of the observer. Most exits the drop at B. Of course, if a much larger amount of light was internally reflected within the drop, we be risking our eyesight if we looked at a rainbow

But when all is said and done, we must accept the evidence of our eyes and acknowledge that, whatever the source of illumination, the colours we see in even the brightest rainbow never matches the intensity of those that Newton claimed are present in a spectrum due to a prism. Indeed, a colourimetric analysis of the hues present in photographs of bright rainbows reveals that the range and purity of colours in a rainbow is a miniscule fragment of what is potentially visible to the naked eye.[16] In the first place, almost all the light that enters a drop passes straight through it. Only a fraction is internally reflected and emerges to form the rainbow arc (Fig. 10.5). At the same time, as Newton recognised, the Sun is not a point source, so the rays responsible for each colour spread out and overlap slightly with their neighbours on emerging from a drop. The combined effect reduces both the brightness and purity of each colour. Other factors that influence the appearance of a rainbow are the size of the drops and the depth of the rain shower in which it is seen. You can easily confirm these facts with a garden spray because you can alter both the depth of the spray and the size of the drops.[17]

Although the rainbow has been an icon of colour since Aristotle's time, fogbows and cloudbows notwithstanding, the number and order of colours said to be visible in a rainbow and, indeed, the very colours themselves has varied from one era to another. Even allowing for the fact that no two rainbows are absolutely identical, surely the colours in a rainbow are there for

---

[16] Lee, R.L., Fraser, A.B. (2001), p. 260–8.

[17] See appendix for details of how to do this.

everyone to see? Not so, apparently. As we saw in chapter three, little is made of the colours of a rainbow in the myths in which it features. The closest Homer comes to referring to its colours in the Iliad is when he compares the iridescence of the enamel on Atreides' breastplate to the rainbow.[18] Elsewhere in the Iliad, he describes the rainbow as being dark or sombre.[19]

In fact, there is a very good reason why the colours of the rainbow are ignored in myths because for most of human history people were largely uninterested in colour for its own sake. Indeed, anthropologists and linguists have found many languages, ancient and current, that don't even have a general word for colour, i.e. a word that refers to hue, let alone names for individual colours.

As for the rainbow, the first unambiguous reference to its colours is to be found in the works of a Greek poet, Xenophanes of Colophon (c. 570 to c. 475 BC). As with all the Greek thinkers of that time, we have only fragments of his work. One of these concerns the rainbow: "She whom men call Iris, this also is by nature cloud, violet, red and pale green to behold."[20] Xenophanes was not a physiologoi, but he was clearly influenced by the Milesian school of Thales, Anaximander and Anaximenes, because in all the surviving fragments of his work that deal with the natural world he seeks physical explanations rather than supernatural ones.

But the first Greek natural philosopher to write extensively on the subject of colour and of the colours of the rainbow was Aristotle. As we saw in chapter four, he claimed that there were only three colours in a rainbow: phoinikous (red), prasinos (green) and halourgos (purple). He dismissed xanthos (yellow) as an illusion due to contrast "for red in contrast to green appears light".[21] In the early fourteenth century Theodoric of Freiberg added a fourth, yellow, and some three centuries later Newton famously expanded their number to seven for reasons that had more to do with Pythagorean numerology than observation. Yet, to judge from how the rainbow is usually represented in art and literature, Newton's seven-coloured rainbow has become firmly lodged in popular imagination.

But, as we have noted in this and previous chapters, a rainbow's colours are at best insipid versions of the intense saturated colours that can be seen in a spectrum of sunlight. Moreover, there are occasions when a rainbow is colourless (e.g. a fogbow). And how do the distinctly non-spectral colours of the supernumerary arcs fit into Newton's spectrum? So, how many colours

---

[18] Homer (1950), XI, 26–28, p. 197.
[19] Homer (1950), XVII, 547–8, p. 330.
[20] Xenophanes of Colophon (1992), pp. 136–9.
[21] Aristotle (1952), chap. II, p. 263.

are there in a rainbow? And why those and not others? Could a rainbow have more colours or fewer? These questions take us to the heart of the nature of the perception of colour.

The apparent indifference to colour in pre-scientific societies is surprising and indeed unexpected. Experience suggests that the perception of colour is automatic and that no special effort is necessary to see colours. Nor is it: even someone who is colour blind can see colours, though far fewer than someone who is not colour blind. But noticing colours is not automatic, if by that is meant being interested in or indeed specifically aware of them in every situation. Although the modern world takes an awareness of colour for granted, cultural considerations turn out to be just as important in the perception of colour as physiology.

Sometime during 1850's William Gladstone (1809–1898), the British statesman and an eminent classical scholar, reached an extraordinary conclusion: the Ancient Greeks must have been colour blind. How else to explain something that he had noticed while studying Homer: Homer seldom used specific names for colours. More tellingly, said Gladstone, Homer often used the same word "…to denote, not only different hues or tints of the same colour, but colours which, according to us are essentially different."[22] Even when allowance is made for the total absence in Homer's time of the vivid artificial colours of our own era, it seemed likely "…that the organ of colour and its impressions were but partially developed among the Greeks of the heroic age."[23]

Gladstone's claim seems to have made little immediate impression on his contemporaries. But it inspired a gifted German philologist, Lazarus Geiger (1829–1870), to study the use of colour words in other ancient texts such as the *Vedas* of ancient India, the *Zendavesta* of the ancient Zoroastrians and the *Hebrew Bible*. In every case he found the same lack of interest in colour as Gladstone had in Homer. Geiger realised that ancient texts could be considered as a fossil record from which the evolution of our current colour vocabulary could be reconstructed. He concluded that this vocabulary had evolved according to a fixed sequence beginning with black, followed by red, yellow and green, with blue being a very late comer indeed.[24]

In more recent times, on several occasions, anthropologists investigating the language of remote and isolated peoples have been taken aback to discover that more often than not their subjects have few if any specific words for individual colours. A particularly ironic episode of thwarted expectation occurred

---

[22] Gladstone, W.E. (1858), p. 458.
[23] Gladstone, W.E. (1858), p. 488.
[24] Deutscher, G. (2010), p. 44.

several decades ago when a small group of Danish anthropologists waded ashore on Bellona, a remote Pacific island, armed with samples of colours for which they wanted the names in Bellonese only to be informed that "we don't talk much about colour here".[25] Nonplussed, the researchers pressed on with their study and found that what words for colour the islander's language possessed appeared to refer primarily to the texture and moistness or freshness of vegetation. Here was direct evidence that not every culture considers that colour in itself is quality worth noting. An earlier study of a Filipino tribe, the Hanunóo, had thrown up broadly similar results.[26]

But in 1969 the results of a detailed survey of words for colour in some one hundred different languages was published by an anthropologist and a linguist that seemed to show that every language has at least two words that are used to name colours and that the vocabulary used to identify colours conforms to a universal pattern that evolves in a seemingly predictable manner.[27] Geiger had concluded as much a century earlier, but his research seems to have been forgotten in the intervening period.

Much of the evidence for this hypothesis was obtained from earlier anthropological studies of isolated societies in New Guinea, Africa and Australia. The remaining evidence was obtained by asking groups of people of different nationalities or ethnicities to select from an array of coloured chips spanning the visible spectrum those that matched most closely the names of colours in their native language. In this way a bilingual dictionary of colour words was assembled that could be used to translate between English and the language being studied. Only names for colours that have a unique reference were accepted, i.e., names not defined in terms of other colours. Moreover, the name had to be a general term for a particular hue; hence, for example, "red" but not "scarlet" or "crimson" or "vermillion".

Brent Berlin (b 1936) and Paul Kay (b 1934), the authors of the study, claimed that their survey established that although not every language has a full range of names for colours, all languages distinguish between lightness and darkness, which they took to be colour words,. More interestingly, any language that has three distinct colour words invariably has a word for red and if it has four colour words, the fourth is either yellow or green. Any language that has five colour words, has words for both yellow and green and if it has six colour words, the sixth is usually blue. Further colour words are added sequentially until a maximum of eleven distinct colour words is reached. These are, in addition to words that distinguish between dark and

[25] Kuschel, R., Monberg, T. (1974), pp. 213–42.
[26] Conklin, H.C. (1955), pp. 339–44.
[27] Berlin, B., Kay, p. (1969).

light (i.e. black and white), red, green, yellow, blue, brown, purple, pink, orange, and grey. No language with between the first three and five colours in this list had a word for blue.

Disappointingly, the authors of the study didn't offer an explanation why colour vocabulary evolves in this manner.[28] One supposes that such an explanation would have to be based on physiological factors rather than cultural ones if only because all humans share the same physiology. But it's difficult to see how physiology alone could determine the evolution of a vocabulary for colour according to a particular sequence (from red to green to yellow to blue etc.). As we shall see, the first six colour words in this scheme are those for primary colour sensations of the eye. But why or how a colour vocabulary should add a distinct word for these colours in the order that is suggested by the Berlin and Kay hypothesis remains an open question.

Why words for colour were coined in the first place seems to be easier to explain. Indeed, Gladstone's idea that naming colours was brought about by the practical needs of painters, dyers and alchemists among others is probably on the right lines.

But Berlin and Kay's research was based on a questionable premise: the assumption that the word used by the subject when asked to put a name to the colour of these chips was primarily the name of the hue. As the study of the vocabularies of ancient Greek, Hanunóo and Bellonese show, there are languages in which the primary use of so-called colour words is not to identify colours but to ascribe properties to things based in some way on their apparent state.[29]

The language of the Hanunóo doesn't even have a word for colour, though it does appear to have four words that in the Berlin and Kay scheme correspond to black, white, red and green. But without a generic word for colour, one cannot ask in Hanuóo: "what colour is that object?". Furthermore, like Bellonese, Hanunóo words for colours are not used to distinguish hues, but rather qualities such as texture or condition. For example, the primary use of the equivalents of our words for red and green in Hanunóo is to distinguish between dryness and wetness (or desiccation and freshness). At the time of the anthropologist's study, the Hanunóo seemed not to have any interest in identifying colours as colours.[30]

---

[28] Berlin, B., Kay, p. (1969), p. 109.

[29] Indeed, this is true of extant languages. In English, green can describe immaturity or inexperience, blue a mood, red anger and yellow cowardice. But the association between colours and, say moods, is not the same in every language that is currently spoken, something you can check using Google Translate.

[30] Conklin, H.C. (1955).

The ancient Greeks did have a word for colour, chroma, but they used their "colour" words to discriminate primarily between brightness and darkness rather than colour. The Greek spectrum was thus arranged in terms of brightness, from leucon (brightness) to melan (darkness). Hence both the sun and water are leucon. Moreover, as in Hanunóo and Bellonese, chloros, the Greek word for green, was used to describe honey, dew, tears and even blood, so its primary reference was to moistness or freshness rather than colour.

Even English once lacked specific words for hues. The colour vocabulary of Old English, which was spoken in Anglo Saxon England between 600 and 1150 AD, was concerned primarily with brightness rather than hue. The only words that referred unambiguously to hue were red and green. As for supposed Anglo Saxon colour words such as glaed, which was used both for bright & shining as well as joyful and happy, only the latter meaning has survived.[31]

Despite an absence of specific colour words in the languages studied by anthropologists, the speakers of those languages had little difficulty in distinguishing colours when tested. One of the first anthropologists to carry out a field study of colour perception, W .H. R. Rivers (1864–1922), discovered that his subjects were not colour blind and could distinguish between colours and shades of colour almost as well as he could.[32]

Whichever view one accepts, the inescapable conclusion is that the world has become colourful principally because we have chosen to make it so. The evidence is that we have singled out different hues by giving them unambiguous names. Cultures that do not have specific words for particular hues appear to have little interest in colour. So it could be said that we have coloured our world using words rather than pigments.

And there is perhaps no better example of this than the rainbow itself. Newton saddled us with the belief that there are seven colours in a rainbow even though you won't always see the full range in every rainbow. But, then again, why bother looking when, apparently, you have Newton's word for it?[33]

Newton, as we saw in chapter six, vehemently maintained that no less than seven distinct spectral colours are necessary to produce whiteness. Yet, by associating colour with refraction, he was forced to concede that there are "as many Degrees of Colours, as there are sorts of Rays differing in

---

[31] Casson R.W., Gardner P.M. (1992), p. 395.

[32] Deutscher, G. (2010), p. 69.

[33] Look up the colours in a rainbow in different European languages and you will find they all list Newton's seven cardinal colours, including that illusive hue, indigo. The rot has spread!

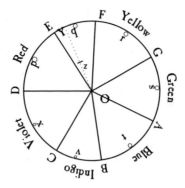

Fig. 10.6 Newton's colour wheel

Refrangibility".[34] We don't perceive an infinite number of distinct colours in the spectrum of sunlight, however, we see five or six. At the same time he was aware that "a mixture of Yellow and Blew makes Green; of Red and Yellow makes Orange; of Orange and Yellowish green makes yellow."[35] He even grudgingly acknowledged, when challenged by Huygens, that blue and yellow could produce a shade of white, though not, he insisted, the white of sunlight.[36]

This gave him the idea for a rather cumbersome method for predicting the result of mixing two or more spectral colours: the colour wheel. This consists of a carboard disc divided into seven segments according to the Pythagorean musical ratios favoured by Newton, each representing one of spectral colours (Fig. 10.6). To find the colour that will be created by mixing two or more colours, small lead weights proportional to the amount of each spectral colour in the mixture are hung from the edge of the circle at the mid-point of each of the relevant segments and the resulting new centre of gravity of the loaded disc is found by trial and error (i.e., by balancing it on a needle). The location of the centre of gravity on the colour circle shows the resultant colour. Newton pointed out that this rather a rough and ready method is "accurate enough for practise, though not mathematically accurate."[37]

At first sight it appears that Newton's colour circle allows for white light to be created by combining two colours. And though this seems possible according to his instructions for using the device, yet "… if only two of the primary Colours which in the Circle are opposite to one another be mixed

---

[34] Newton, I. (1952), p 122.

[35] Newton, I., (1671/2 b), p. 3082.

[36] Newton, I. (1673), p. 6087.

[37] Newton, I. (1952), p. 158.

in an equal proportion, the point Z shall fall upon the center 0, and yet the colour compounded of those two shall not be perfectly white, but some, faint anonymous Colour. For I could never yet by mixing only two primary Colours produce a perfect white."[38]

If it is possible to combine two colours to form a third one, however, is it not likely that white can be produced from fewer colours than the Newtonian septet? The answer was provided indirectly by the principle of trichromacy adopted by natural philosophers, painters and printmakers in the eighteenth century. This is the idea that just three primitive colours—red, yellow and blue—can be mixed to make all other colours. The first person to state this idea unequivocally was Edmé Mariotte in 1681: "There are five main ones: white, black, red, yellow & blue: All the others can be made by mixing some of these: yellow & blue mixed together are green, red and blue are violet."[39]

No distinction, however, was made between pigments and lights; it was assumed that what was true of mixing pigments applied to coloured lights. Newton himself made no distinction between pigments and lights, and claimed that he had succeeded "to compound a white, mixing coloured Powderes which Painters use [from] Orpiment [yellow], Purple, Bise [deep blue] and Viride Aeris [grayish green]."[40] The result was actually grey, but at the time it satisfied him.

As to the cause of coloured light, Newton answer was "For the Rays to speak properly are not coloured. In them there is nothing else than a certain Power and Disposition to stir up a Sensation of this or that Colour."[41] But the implication of this important distinction was not fully grasped until 1802 when Thomas Young realised that colour is a matter of physiology not physics, and sketched out a feasible explanation of colour perception. He realised that if "the Rays ... are not coloured" then trichromacy is a consequence of the workings of the visual system. He pointed out that it is not necessary for the retina to be sensitive to every individual colour in the visible spectrum, only to three primary colours.[42] He initially accepted the established trichromatic triplet of red, yellow and blue, but later changed this to red, green and violet. The advantage of the latter triplet, in his view, is that combining red and green lights makes yellow and green and violet makes blue.[43] Hence sensitivity to light rays responsible for sensations of red, green

[38] Newton, I. (1952), p. 156.

[39] Mariotte, E. (1717), p. 282.

[40] Newton, I, (1952), pp. 150–4.

[41] Newton, I. (1952), p. 125.

[42] Young, T. (1802a), pp. 20–1.

[43] Young, T. (1802b), p. 395.

and violet would suffice to account for all the colours seen in the visible spectrum. This colour triplet also explains why we can see non-spectral colours such as magenta, a combination of red and blue.

Young's trichromatic theory of vision was met with indifference until taken up several decades later, first by Herman von Helmholtz in 1852 and later by James Clerk Maxwell in 1855. Helmholtz was at first reluctant to accept Young's hypothesis because he was unable to match the colours in the spectrum by combining individual coloured lights, but came around in 1858 when he refined his colour mixing technique.[44] The existence of three types of retinal cell, each sensitive to a broad range of the spectrum was confirmed by Maxwell, who determined the shapes of the response curves of the three colour-sensitive receptors.[45] Between them they confirmed that "the Rays … are not coloured" and that colour is a matter of physiology not physics,.

Although Helmholtz's 1852 colour mixing experiments seemed initially to contradict Young, they established clearly that there is a fundamental difference between mixing lights and mixing pigments. Mixing lights is additive because combining them makes the result brighter. White light is brighter than the individual colours that make up its spectrum. The colour of a pigment, on the other hand, is due to which parts of the spectrum of white light it absorbs, hence it is always darker than the light illuminating it. A red ball looks red because its surface pigment absorbs most of the remainder of the spectrum, namely the green to violet portion.

Painters, of course, had long known that combining pigments creates new colours at the expense of brightness. Helmholtz's discovery seemed to provide a solution: don't mix pigments, keep them separate to maintain their brightness. And the painter who first exploited this advice was George Seurat (1859–1891), the inventor of pointillism.[46] His idea was to preserve the purity and brightness of pigments by applying them as individual dots, which when viewed from a suitable distance merge together additively to produce broad fields of colour.

The assumption that there is no difference between subtractive pigments and additive lights is charmingly illustrated in Angelica Kauffman's (1741–1807) painting, *Colour*.[47] It shows a painter using her brush to transfer colours from a pale, emblematic rainbow to her empty pallet, colours she will

---

[44] MacAdam, D.L. (1970), pp. 84–100.

[45] Mahon, B. (2003), pp. 49–56.

[46] Kemp, M. (1990), p. 312.

[47] Kauffman, A., (1778–80), *Colour*, Royal Academy, London: https://tinyurl.com/2p85783a. Accessed 31/07/22.

use to paint her rainbow. But as painters have found, capturing the rainbow's diaphanous coloured light in paint is difficult.

Quite apart from colour, the challenge that rainbows pose to landscape painters is that, unlike, say, clouds or trees, rainbows are both infrequent and transient. Painters will have few opportunities to thoroughly familiarise themselves with rainbows in the open air and so often fall back on the clichéd semicircular arc of seven distinct colours. Moreover, a rainbow has a precise geometrical relationship with the sun and the observer so its correct placement in a landscape depends on the direction from which the sun illuminates the scene. Representing a rainbow accurately and convincingly on canvas is constrained by season, time of day and direction from which the scene is illuminated by the sun, to say nothing of the painter's skill and pallet. Yet, for all that, the rainbow is the only atmospheric arc that has consistently attracted the attention of painters.

An artist unfamiliar with rainbows and their features will inevitably get some of its details wrong, resulting in a lurid arc placed incorrectly within a landscape. Unlike depictions of clouds, those of rainbows unsparingly expose artifice. Its earliest pictorial representations in Western art were allegorical and usually inspired by the story of Noah and the Flood. Later painters took the rainbow as a symbol of Christ's majesty and transcendence with little regard to its real appearance, drawing on whatever ideas about the symbolism of light and colour were current at the time rather than on observation.

This may not have mattered when rainbows were included in paintings merely as religious symbols, but the advent of naturalism in the late sixteenth century encouraged painters to try their hand at adding a rainbow to a pastoral scene. Yet few succeeded in placing them correctly, let alone capturing their ephemeral translucent colours. The magnificent rainbow that arches over the landscape in Peter Paul Rubens' (1577–1640) celebrated *The Rainbow Landscape*[48] is wrong in several respects, the most obvious of which is that it is at an angle to the plane of the picture and looks as if it were a solid object.[49] Misplacing a rainbow in a landscape painting is a common mistake. Seurat added a partial rainbow as an afterthought to a preliminary study for his *Bathers at Asnières*.[50] It wasn't included in the final work, which is just as well because in the study the rainbow is at right angles to the shadows in the picture, whereas it should be perpendicular to them.

---

[48] Rubens, P.P., c. 1636, *The Rainbow Landscape*, The Wallace Collection, London: https://tinyurl.com/3keays8b. Accessed 28/07/22.

[49] Lee, R.L., Fraser, A.B., 2001, p. 125.

[50] Seurat, G., 1883, *Study for Bathers at Asnières*, The National Gallery, London: https://tinyurl.com/ycx85zku. Accessed 28/07/22.

Arguably, of all painters, the most assiduous student of rainbows was John Constable (1776–1837). He considered that "Painting is a science, and should be pursued as an enquiry into the laws of nature. Why then may not landscape painting be considered as a branch of natural philosophy, of which pictures are but the experiments?"[51] He lived up to this where his cloud studies are concerned but often bent the rules with rainbows. Nowhere more so than in his 1831 *Salisbury Cathedral from the Meadows*.[52] The direction from which the menacing cloudscape and the West end of the cathedral is illuminated is at odds with the rainbow that arches over the cathedral. The sun is clearly out of frame to the right of the scene. Constable knew that the arc of a rainbow is always perpendicular to the direction of the sun, so it could not form where he painted it. Moreover, the length of shadows of the tree stumps in the foreground indicate that the sun is some 65° above the horizon, and thus far too high for a rainbow to form. The only possible reason for placing a rainbow in the scene in the position it occupies is as a sign of hope in contrast to the threatening sky, which seems to have been Constable's intention.

For a painter, Constable was extremely knowledgeable about rainbows, as his notes on the subject attest.[53] But in spite of his belief that painting should be an enquiry into the laws of nature, he was prepared to set aside that knowledge when it clashed with his artistic aims, as his Salisbury painting demonstrates. Later that year he painted a watercolour, *View from Hampstead, with a Double Rainbow"* executed "between 6. & 7. oclock/Evening June 1831,[54] that captures a rare example of a section of the primary and secondary arcs formed within a narrowing beam of sunlight known as an anticrepuscular ray. Although the relative widths of the arcs and the gap between them are far from accurate—the secondary arc should be twice as broad as the primary, but is only 25% wider—the painting is otherwise meteorologically correct. Which only goes to show how difficult it is to capture a rainbow in paint on the fly.

Another of Constable's works, *Stoke by Nayland, Suffolk*, an 1829 print of an earlier 1810 work in oil, adds a rainbow to the original oil painting.[55] It depicts, according to Constable, "the solemn stillness of Nature in a Summer's

[51] Leslie, C.R., 1912, p. 285.

[52] Constable, J., 1831, *Salisbury Cathedral from the Meadows*, Tate Britain, London: https://tinyurl.com/yn9kmvt6. Accessed 28/07/22.

[53] Thornes, J.E. (1999), pp. 85–8.

[54] Constable, J. (1831), *View from Hampstead, with a Double Rainbow*, British Museum, London: https://tinyurl.com/25zb8s8u. Accessed 28/07/22.

[55] Constable, J. (1829), *Stoke by Nayland*, Victoria and Albert Museum, London: https://tinyurl.com/4uf4z6kt. Accessed 28/07/22.

Noon." Yet he knew perfectly well that at midday rainbows are not possible in summer, so its purpose is merely picturesque.

Constable is by no means unusual in ignoring the reality of rainbows for artistic effect. A review of some nineteenth century paintings by major artists that include a rainbow, Joseph Wright's *Landscape with a Rainbow*,[56] W.M.Turner's 1798 *Buttermere Lake*[57] and 1807 The *Wreck Buoy*[58], Frederic Edwin Church's 1857 *Niagra*[59] and 1866 *Rainy Season in the Tropics*,[60] John Everett Millais 1856 *The Blind Girl*,[61] all suggest that painters have been more attracted by the rainbow's picturesque qualities and symbolism than as a mere meteorological phenomenon within a landscape.

In the years following his death in 1727, Newton's ideas on light and colour were enthusiastically embraced by poets, who did much to introduce his ideas on light and colour to the masses. Enlightened by reading *Opticks*, the rainbow became a favourite poetic motif, not least because they saw it as a vindication of Newton's claim that white light is an amalgam of prismatic colours. These Newtonian poets, as they have been called, came to think of light as the source of beauty because it was the source of colour. Among the poets who championed Newtonian optics, was the Scottish poet, James Thomson (1700–48), who wrote *The Seasons*, a four-part meditation on the natural world, albeit from deist slant. A Newtonian take on the rainbow appears in "Spring".

Mean time refracted from yon eastern cloud,
Bestriding earth, the grand æthereal bow
Shoots up immense! and every hue unfolds,
In fair proportion, running from the red,
To where the violet fades into the sky.
Here, mighty Newton, the dissolving clouds
Are, as they scatter'd round, thy numerous prism,
Untwisting to the philosophic eye

[56] Wright, J. (1793), *Landscape with a Rainbow*, Derby Museum, Derby: https://tinyurl.com/s3s4t4h4. Accessed 28/07/22.

[57] Turner, W.M. (1798*), Buttermere Lake*, Tate Britain, London: https://tinyurl.com/2vkmfh76. Accessed 28/07/22.

[58] Turner, W.M. (1707), *The Wreck Buoy*, Walker Art Gallery, Liverpool: https://tinyurl.com/38kvz5v8. Accessed 28/07/22.

[59] Church, F.E. (1857), *Niagra*, National Gallery of Art, Washington, USA: https://tinyurl.com/3939ht8n. Accessed 28/07/22.

[60] Church, F.E. (1866), *Rainy Season in the Tropics*, Fine Arts Museums of San Francisco, USA: https://tinyurl.com/5xa84rbr. Accessed 28/07/22.

[61] Millais, J.E. (1856), *The Blind Girl*, Birmingham Museums Trust, Birmingham https://tinyurl.com/5daur6fy. Accessed 28/07/22.

The various twine of light, by thee pursu'd
Thro' the white mingling maze.[62]

Thomson's goes on to juxtapose the Newtonian insight into the source of the rainbow's colours with the naïve yokel who sees them but does not understand their cause.

But a later generation of poets either ignored Newton or turned against him. By the turn of the nineteenth century not only were Newton's ideas on light being overturned by Young and Fresnel, poets such as William Blake (1757–1827) and John Keats were vigorously attacking his cultural legacy. Keats in particular held Newton responsible for spoiling the aesthetics of the rainbow with mathematics and experiments.

Newton was not, of course, the author of the materialism they abhorred. The idea that the world is a vast inanimate mechanism had been formulated a generation before Newton, principally by René Descartes. But Newton was by far the most successful practitioner of the mechanical philosophy, as it came to be known, and so attracted a disproportionate degree of censure from those appalled by the idea of the world as a mere machine devoid of purpose, beauty or meaning. Moreover, Newton was not quite the hard headed materialist they took him for because not only did he assume that Pythagorean numerology underpinned the spectrum of sunlight, he also believed that the application of his theory of gravity to the motion of the planets revealed the hand of God at work.

During a dinner party held in 1817 in the studio of the painter Benjamin Haydon (1786–1846), Newton was mocked by Keats, among others, as "…a fellow who believed nothing unless it was as clear as the three sides of a triangle" who had "destroyed all the poetry of the rainbow by reducing it to the prismatic colours."[63] Three years later Keats voiced his dismay at the materialist vision of nature in these, by now, well-known lines:

Do not all charms fly
At the mere touch of cold philosophy?
There was an awful rainbow once in heaven:
We know her woof, her texture; she is given
In the dull catalogue of common things.
Philosophy will clip an Angel's wings,
Conquer all mysteries by rule and line,

---

[62] Thomson, J. (1908), p. 11.
[63] Haydon, B.R. (1926), p. 269.

Empty the haunted air, and gnomed mine-
Unweave a rainbow...[64]

Keats has been taken to task for his rejection of scientific accounts of the
rainbow.[65] His claim that science robs the rainbow of beauty and meaning
by reducing it to an inanimate event to which science alone holds the key has
been criticised as narrow-minded and wrong-headed. If we have lost a sense of
the rainbow's other-worldliness, science has given us something in return by
opening our eyes to its variety and complexity. Constable, a contemporary of
Keats, believed that "We see nothing truly till we understand it."[66] That's why
he made a detailed study of the optics of rainbows and the meteorology of
clouds so that he could capture them as accurately as possible on canvas. Did
Keats ever consult the dull catalogue of science on the subject of the rainbow?
Or was his mind closed to the possibility that it might have something to offer
the poet, as it did to Constable, the painter?

Perhaps we shouldn't rush to judgement. Isn't a spontaneous, untutored
enjoyment of a rainbow preferable to one that holds it at arms-length with
impersonal scientific theories of light and colour? There are those who would
unhesitatingly agree with Keats when wrote "O for Life of Sensations rather
than of Thoughts!"[67] and agree that the scientific account of nature dulls
the imagination and replaces an appreciation of its otherworldly beauty with
leaden mechanical explanations. John Ruskin (1819–1900), the nineteenth
century English writer and art critic, suggested that the pleasure and wonder
that an unlettered peasant might feel at the sight of a rainbow was greater than
that of someone who understood its underlying optical principles. "For most
men, an ignorant enjoyment is better than an informed one … I much question whether any one who knows optics, however religious he may be, can
feel in equal degree the pleasure or the reverence which an unlettered peasant
may feel at the sight of a rainbow."[68] What arrant nonsense. He obviously
had not read James Thomson on the subject of yokels and rainbows.

For all that, we shouldn't avoid asking whether the rainbow can retain its
freshness for someone familiar with its scientific explanation. Isn't it possible
that the sight of a rainbow becomes increasingly prosaic and unengaging as
one's understanding of it increases? That might be true if its appearance never
varied.

---

[64] Keats, J. (1909), p. 41.
[65] Dawkins, R. (1998), pp. 38–65.
[66] Leslie, C.R. (1896), p. 393.
[67] Keats, J. (1952), p. 67.
[68] Ruskin, J. (1906), p. 293.

In Keats' defence, it has to be said that Newton's account of the rainbow, the most complete scientific explanation of the rainbow known in Keats' day, is highly idealised, based as it is on the assumption that a combination of refraction and reflection alone is responsible for its shape and colours, and that all raindrops are spherical. It ignores the most important factor that determines the colours seen in a bow, as well as the width and diameter of the arc: the size and shape of raindrops. As we noted in chapter seven, not all drops are spherical, and not all spherical drops are sufficiently large for refraction to occur unaccompanied by noticeable diffraction and interference.

If Newton's account had been the last word on the subject, and a rainbow is due solely to refraction and reflection within large spherical drops, then it would be a far less interesting and varied phenomenon than it is. The explanation of his idealised bow can be grasped without much difficulty by anyone with an elementary understanding of optics. Yet once mastered, its novelty might well pale, since, except for its brightness, which depends on factors such as the size of the raindrops, the Sun's brightness and the background against which the rainbow is seen, its shape, size and colours would be exactly the same from one occasion to the next. A rainbow would still be a uplifting spectacle, but of the "seen one, seen them all" variety, not a sight to seek out time and again and linger over its finer details in the hope of seeing one or other of its rarer features.

The Newtonian rainbow is cut and dried and, dare one say, rather dull. It suggests that there are no surprises in store, no subtle variations in colour from one bow to another and, of course, no supernumerary arcs. One of the drawbacks of Newton's account is that it seemed to confirm what was already known. It renders ordinary what is in reality an extraordinarily mixed bag of optical effects. People had been looking at rainbows for thousands of years, and its major features had been entered in science's supposedly dull catalogue, something that, as records show, inhibited close observation. If one bow is much like another, there's a temptation to forgo observation in favour of theory. What's the point in searching out rainbows in order to learn more about the phenomenon if the scientific explanation has already done that for you, particularly if the explanation suggests that all rainbows are necessarily alike? And if this is so, perhaps you might tempted to agree with Keats, and find the scientist's rainbow ordinary and dull in comparison to that of the poet, who, ignorant or indifferent to what Newton had to say on the subject, might well claim to have seen and been moved by rainbows unimagined by science.

Happily for those of us intrigued by rainbows, Newton didn't have the last word. As we saw in previous chapters, light has several tricks up its sleeve that

Newton didn't fully understand and which make rainbows endlessly varied. The appeal of a rainbow lies as much in details that are subtly different from bow to bow, as in its overall appearance, a huge luminous arc of many colours arching above a landscape. Its mutability ensures that there are treats in store for those in the know almost every time one sees a rainbow.

Scientific accounts of the rainbow usually concentrate on the role of sunshine and raindrops at the expense of the eye. The eye is regarded merely as a passive terminus for the reflected and refracted light that emerges from a multitude of raindrops. The complex act of seeing is taken for granted, something that unwittingly places all sighted creatures on the same footing. Yet there is no evidence that birds, creatures with the finest eyesight in nature, are aware of rainbows. Birds, of course, are more likely to see rainbows than earthbound beings because when on the wing they are ideally placed to see rainbows, circular ones at that, whenever they fly near sunlit rain.

A bird may see a rainbow in the sense that an image is formed on the retina within its eye, but does it notice it? Something more than an eye is required, something that, as far as we can tell, only humans possess: the ability to invest experience with meaning that goes beyond brute perception and immediate biological need. The vehicle of meaning, not just of what we say, but of what we do and see and think is language. Even if birds noticed rainbows, how would we know? Birdsong, for all its delightful complexity, is not language, any more than are our sobs, groans and laughter. Even allowing for the fact that birdsong is meaningful to birds, it is not capable of handling or conveying ideas. A bird can't chirp about the colours in the primary bow, or be aware that it can see colours in the rainbow that we can't. Birdsong serves more immediate needs, such as marking out territory or attracting a mate.

Colour perception is a far more complex process than the earlier investigators such as Young, Maxwell and Helmholtz took it to be. And, as we noted in connection with words for colour, arguably, culture is as important to perception as physiology. We now know that humans are unusual among mammals in having colour vision. Yet human colour vision is not as broad as that of birds because in addition to visible light they can see ultraviolet light, which we can't. But the limitations of human colour vision compared to that of birds is more than compensated for by language, which has allowed us to go as far beyond the boundaries of brute perception as our wit and imagination allow. Language, after all, has enabled us to discover that birds see colours we can't.

It's tempting to assume that language is simply a means to describe the world, that words are by and large the names of things for which they stand.

But although naming is an important feature of language, it could be said that this is the least important of its functions. Language enables us to make links between things that, on the face of it, seem to have nothing in common. Thus it is that the rainbow has always been more than an optical spectacle. It is such a striking sight that it has taken on meanings and associations that make no objective sense because it has several attributes that strike a chord with us and which inform its significance in mythology. Foremost among these is its ancient role as a bridge between worlds. And in recent times people have found in its many colours a useful metaphor for unity in diversity.

Hence the several rainbow coalitions in politics that unite people of different cultures and ethnicities. Desmond Tutu (1931–2022), Archbishop of Cape Town, coined the phrase "rainbow nation" in 1994, drawing both on the rainbow's significance to Christians as well as the presence of many colours within the arc that have a common origin in light, to describe the multiracial and multicultural democracy that emerged in South Africa from the dark years of apartheid.

Rainbow flags, on the other hand, appear to have little to do with rainbows apart from being multicoloured banners that employ Newton's canonical colours which make them sufficiently distinctive to stand out from all others. The best known of these is the rainbow flag created by Gilbert Baker (1951–2017) in 1978 as a symbol of the gay community in San Francisco. It has since been adopted by the LGBT community and is now widely accepted as a symbol of sexual diversity by the public at large. The original flag had eight colours, but within a year of its creation was reduced to six: red (symbolising life), orange (healing), yellow (sunlight), green (nature), blue (harmony), violet (spirit). Pink (sexuality) and turquoise (art/magic) were removed for practical reasons to do with manufacture.

Drawing on the biblical interpretation of the rainbow as a symbol of peace and harmony, the rainbow flag has been adopted by various peace movements. It was first deployed during a peace march in Italy in 1961 and later used in 2002 during the "Pace da tutti i balcony" ("peace from every balcony") campaign as a protest against the imminent war in Iraq.

As we saw in chapter three, the Inca's flag included a rainbow, but whether this is the source of modern rainbow flags in Peru and Bolivia remains an open question. Nor can one be sure of the inspiration for the hand-drawn rainbows that appeared in windows during the worst period of the Covid-19 pandemic. It's tempting to surmise that the idea that a rainbow represents hope in an uncertain time is likely to have been the source.

One way or another, the rainbow is as close to a universal symbol as any I can think of, one that appears always to have appealed to people of different cultures and interests—as interesting and fascinating to the scientist as it is to the poet and the painter, not to mention the casual bystander. It's a phenomenon of which people are unlikely ever to tire, an inexhaustible source of wonder and fascination. And this book is unlikely to be the last word on the subject. "Knowing another is endless ... The thing to be known grows with the knowing."[69]

---

[69] Shepherd, N. (2011), p. 108.

Appendix

Twenty-Four Rainbows to See Before You Die

1. Primary rainbow in the afternoon (i.e.in the eastern sky)
2. Primary rainbow in the morning (i.e.in the western sky)
3. Primary rainbow at midday (i.e.in the northern sky)
4. Primary rainbow at sunset (i.e. a 180° semicircle)
5. Primary rainbow more than 180° or completely circular
6. Primary rainbow segments (rainbow wheel)
7. Primary and secondary bow (i.e. a double bow)
8. Tertiary rainbow
9. Primary and secondary bow with pronounced dark band
10. Supernumerary bows to primary bow
11. Supernumerary bows to secondary bow
12. Moonbow
13. Dewbow
14. Fogbow
15. Cloudbow
16. Twinned bows
17. Reflection of a rainbow in water
18. Reflected light rainbow
19. Glass bead rainbow
20. Searchlight bow
21. Garden hose spraybow
22. Spraybow in a fountain

© The Editor(s) (if applicable) and The Author(s), under exclusive
license to Springer Nature Switzerland AG 2023
J. Naylor, *The Riddle of the Rainbow*, Copernicus Books,
https://doi.org/10.1007/978-3-031-23908-3

23. Spraybow in a waterfall
24. Spraybow in seawater.

Tips on Photographing a Rainbow

1. You will need a very wide-angle lens to capture the entire arc of a semicircular rainbow. The angular diameter of the outer (red) edge of the primary rainbow is 84°, that of the outer (violet) edge of the secondary bow is 108°. Should you wish to calculate the field of view of your lens: https://tinyurl.com/26zbs4r2
2. A rainbow reaches its maximum size when the sun is at the horizon. At other times (and seasons) less of the arc is visible, so you won't need a wide-angle lens.
3. Bracket exposures. Slight underexposure will tend to enhance the colours of the rainbow giving deeper saturation.
4. A tripod helps if the rainbow is faint, as so many of them are.
5. Background affects the brightness of a rainbow. A rainbow stands out better against a dark background, e.g. grey storm clouds.
6. Best photographed in flat, open countryside to capture a complete arc.

Some Experiments with Rainbows

1. Spraybows. Use the spray nozzle of a garden hose to create rainbows on a sunny day or on clear night when there is a bright full moon high above the horizon. Experiment with sprays of different drop sizes and notice how very different the colours are in those formed in large drops compared to those formed using the mist setting. Create a circular bow by moving the spray about your shadow.
2. Investigating the colours of the rainbow with Descartes. There is one experiment that anyone who wants to get a better insight into how a rainbow comes about should perform. It's not necessary to use a spherical flask, handy though this is, or a crystal ball. A tall tumbler or large wine glass (clear glass or plastic) will do as long as it has a smooth circular cross section. Fill it with water, go outside and turn your back to the sun. Hold the glass at arms-length within the shadow of your head as near to the antisolar point as possible and tilt it away from you until it is perpendicular to the sun's rays. Gradually move the glass out from your shadow

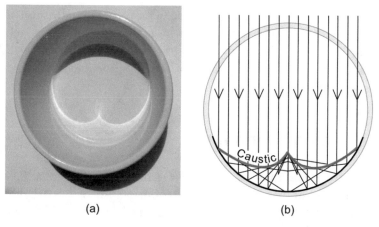

Fig. A.1 a A photograph of a caustic surface due to the reflection of sunlight from the curved surface of a cup. b A ray diagram showing how this caustic surface is created

keeping your eyes fixed on the glass. At some point you will see a brilliant red spot of light just inside its inner edge. This is the rainbow ray responsible for a primary rainbow's red band. Further bright flashes corresponding to the other rainbow colours will appear as the glass moves away from your shadow. Continue moving the glass and another sequence of bright spots will appear near the far edge of the glass which correspond to the secondary bow. The experiment is quite fiddly, so be prepared for trial and error.

Caustic Surfaces

Caustic surfaces due to reflections and refraction of light by curved surfaces are the source of many of the bright spots and streaks of light that we come across in daily life. They come about wherever rays of light cross to create a bright envelope. You will often have noticed a caustic surface at the bottom of a tea cup due to reflection from its side (Fig. A.1). And caustics due to refraction are the source of the bright, stripes seen snaking across the bottom of a pool of clear, rippled water on sunny days. The bright spot produced by focusing sunlight with a convex lens can ignite paper, a particularly impressive example of a caustic surface.[1]

---

[1] For more on caustic surfaces due to refraction see https://tinyurl.com/57sk86yf. Accessed 10/09/22.

The rainbow is a very unusual example of a caustic curve because not only does it involve a concentration of light in a particular direction, refraction also brings about dispersion so that a series of caustic surfaces of different colours appear next to one another. But none bright enough to force one to look away, happily!

Polarisation of Rainbows

Polarised light is common in nature. Nevertheless, we are usually unaware of it because the human eye is unable to distinguish between light that is polarised and light that is not polarised without the use of artificial aids such as the polarising filters fitted to some types of sunglasses.

The simplest way to tell if light is polarised is to look at it through a polarising filter. If there is a change in brightness as you rotate the filter then the beam contains some polarised light. The change in intensity is due to the action of the material of which the filter is made. Only light that is polarised in a particular plane can pass through the filter. Light that is not polarised in this plane is partly or wholly absorbed by the filter. The filter has only to be rotated through a quarter turn to extinguish light that is polarised.

Many people can see a faint butterfly shaped yellow spot when looking at a broad expanse of polarised light such as the zenith sky at midday or a blank computer screen. It is known as *Haidinger's Brush*, after K.R. Haidinger (1795–1871), the Austrian minerologist who first noticed it. The effect is particularly noticeable if you rock your head from side to side.

Since a rainbow involves the reflection of light, you would expect it to be polarised and, indeed it is. In fact, it is one of the most highly polarised coloured arcs in nature because the angle of reflection of the rainbow ray responsible for the primary arc within a drop of water is very close to the Brewster angle of 38°. Hence if you look at a rainbow through a polarising filter, segments of the bow will disappear and reappear as you rotate the filter. Moreover, the orientation of the filter that absorbs light from the apex of the bow will not block light from its sections at an angle to the horizon, proof that a rainbow is tangentially polarised.

References

Abragam, A. (1988). Louis Victor Pierre Raymond de Broglie. *Biographic Memoirs of Fellows of the Royal Society, 34*, 23–41. https://tinyurl.com/ywraed4t.

Airy, G. B. (1838). On the intensity of light in the neighbourhood of a caustic. *Transactions of the Cambridge Philosophical Society, 6*, 379–403.

Al-Haytham, I. (1989). *The optics of Ibn Al-Haytham, Books I—III On direct vision* (A.I. Sabra, Trans.). The Warburg Institute University of London. https://tinyurl.com/bdf9jfb2.

Aldersey-Williams, H. (2020). *Dutch Light, Christiaan Huygens and the making of science in Europe*. Picador.

Andriesse, C. D. (2005). *Huygens: The man behind the principle*. CUP.

Anlezark, D. (2013). *Water and fire, The myth of the Flood in Anglo-Saxon England*. Manchester University Press.

Anon. (1676). A narrative of a strange effect of thunder upon a magnetic sea-card. *Philosophical Transactions of the Royal Society, 11*, 647–653. https://tinyurl.com/2p8bc8fs.

Arago, F. (1857). *Biographies of distinguished scientific men* (W. H. Smyth & others, Trans.). Longman. https://tinyurl.com/zsscvwen.

Arago, F. (1858). *Œuvres complètes de François Arago* (vol. 7). Paris https://tinyurl.com/4ffse9pe.

Arianrhod, R. (2019). *Thomas Harriot, A life in science*. OUP.

Aristotle. (1933). *Metaphysics* (H. Tredennick, Trans.). Loeb Classical Library. https://tinyurl.com/yc4nxaap.

Aristotle. (1952). *Meteorologica* (H. D. P. Lee, Trans.). Loeb Classical Library.

Armitage, A. (1966). *Edmond Halley*. Thomas Nelson.

© The Editor(s) (if applicable) and The Author(s), under exclusive license to Springer Nature Switzerland AG 2023
J. Naylor, *The Riddle of the Rainbow*, Copernicus Books, https://doi.org/10.1007/978-3-031-23908-3

Armitage, A. (1950). Rene descartes (1596–1650) and the early royal society. *Notes and Records: Royal Society of London, 8,* 1–19. https://tinyurl.com/3nv8k86e.

Babbage, C. (1830). *Reflections on the decline of science in England and on some of its causes.* https://tinyurl.com/ykeurh5c.

Babinet, J. (1837). Mémoires d'optique météorologique. *Comptes Rendus, IV,* 638–648. https://tinyurl.com/y7tjj3y3.

Bacon, R. (1897). *Opus Majus* (vol 2). In J.H. Bridges (Ed.). Clarendon Press. https://tinyurl.com/yck3nwau.

Baillet, A. (1691). *La Vie de Monsieur Des-Cartes.* A Paris, Chez Daniel Horthemels. https://tinyurl.com/24pj5ekn.

Barth, M. (1995). Huygens at work: Annotations in his rediscovered personal copy of Hooke's Micrographia. *Annals of Science, 52*(6), 601–613.

Bartholini, E. (1669). *Experimenta crystalli Islandici disdiaclastici, qvibus mira & insolita refractio detegitur.* Hafniae. https://tinyurl.com/27u5xpab.

Bell, A. E. (1948). *Christian Huygens and the development of science in the seventeenth century.* Edward Arnold.

Bence Jones, H. (1870). *The life and letters of Faraday* (vol. 1 & 2). Longmans, Green & Co. https://tinyurl.com/mum5x5xx, https://tinyurl.com/2s43waza.

Berlin, B., & Kay, P. (1969). *Basic color terms: Their universality and evolution.* University of California Press.

Bernard Cohen, I., & Westfall, R. S. (Eds.). (1995). *Newton: Texts, backgrounds, commentaries.* W.W.Norton & Co.

Bernouilli, J. (1742). *Opera Omnia* (vol. 1–4). Lausannae & Genevae.

Berry, M. (2020). True, but not real. *Physics World, 33*(1):56.

Billet, F. (1868). Mémoire sur les dix-neuf premiers arcs-en-ciel de l'eau. *Annales Scientifiques de l'École Normale Supérieure, 5,* 67–109. https://tinyurl.com/yckpepcs.

Biographia Britannica. (1757). *Biographia Britannica, or the lives of the most eminent persons who have flourished in Great Britain and Ireland from the earliest Ages, down to the present times* (vol. 4). https://tinyurl.com/mwhp8s5d.

Biot, J.-P. (1826). *Experimental treatise on optics* (vol. 3). https://tinyurl.com/sxw3aevv.

Biot, J.-P. (1811). Sur la Dissection de la lumiere par des reflexions et des refractions successives, lu à l'Institut par M.Biot, le 11 mars 1811. Moniteur Universel 73:14 Mars. https://tinyurl.com/5ffyerf2.

Birch, T. (1757). *The history of the royal society of London for improving of natural knowledge* (vol. 3). London. https://tinyurl.com/bddew2ws.

Blust, R. (2000). The origin of dragons. *Anthropos, 95*(2), 519–536. https://tinyurl.com/4jk2adjx.

Blust, R. (2021). Pointing, rainbows, and the archaeology of mind. *Anthropos, 116*(1), 145–162. https://tinyurl.com/2p8vvra8.

Boas Hall, M. (1984). All scientists now: The royal society in the nineteenth century. CUP.

Boyer, C. B. (1987). *The rainbow.* Macmillan.

Boyle, R. (1664). *Experiments and considerations touching colours.* https://tinyurl.com/33cbvftw.

Bravais, M. A. (1845). Notice sur les Parhélies situés a la méme hauteur que le soleil. *Journal de L'École Royale Polytechnique, 13.* https://tinyurl.com/2rrss8bp.

Brewster, D. (1813). *A treatise on new philosophical instruments, for various purposes in the arts and sciences with experiments on light and colours.* Edinburgh. https://tinyurl.com/4udtkrf8.

Briggs, R. (1991). The académie royale des sciences and the pursuit of utility. *Past & Present, 131*(1), 38–88.

Burleigh, M. (2005). *Earthly powers.* Harper Collins.

Businger, S. (2021). The secrets of the best rainbows on Earth. *Bulletin of the American Meteorological Society, 102*(2), E338–E350. https://tinyurl.com/3npnca8x.

Campbell, L., & Garnett, W. (1882). *The life of James Clerk Maxwell: With a Selection from his correspondence.* Macmillan and Co. https://tinyurl.com/352eyuab.

Casson, R. W., & Gardner, P. M. (1992). On brightness and color categories: Additional data. *Current Anthropology, 33*(4), 395–399.

Chapman, A. (2004). *England's Leonardo: Robert Hooke and the seventeenth-century scientific revolution.* Taylor & Francis.

Christensen, D. C. (2013). *Hans Christian Oersted, Reading Nature's Mind.* OUP.

Cobo y Peralta, B. (1893). Historia del Nuevo mundo por el Padre Bernabé Cobo de la Compañía de Jesús. Sevilla. https://tinyurl.com/bd72cnzm.

Cohn, N. (1995). *Cosmos, chaos and the world to come.* Yale UP.

Cohn, N. (1996). *Noah's flood: The genesis story in western thought.* Yale UP.

Collingwood, R. G. (1960). *The idea of nature.* OUP.

Conklin, H. C. (1955). Hanunóo color terms. *Southwestern Journal of Anthropology, 11*(4), 339–344.

Corliss, W. R. (1983). *Handbook of unusual natural phenomena.* Anchor Books.

Crombie, A. C. (1953). *Robert grosseteste and the origins of experimental science* (pp. 1100–1700). OUP.

Crossland, M. (1967). *The society of Arcueil; A view of French science at the time of Napoleon I.* Heineman.

Cunningham, A., & Jardine, N. (Eds.). (1990). *Romanticism and the sciences.* CUP.

Dalley, S. (1989). *Myths from Mesopotamia: Creation, the flood, Gilgamesh, and others.* OUP.

Dalzel, A. (1862). *History of the university of Edinburgh from its foundation with a memoir of the author* (vol. 1). Edinburgh. https://tinyurl.com/2p85bcbn.

Darrigol, O. (2012). *A history of optics from Greek antiquity to the nineteenth century.* OUP.

Davidson, I. (2010). *Voltaire.* Profile Books.

Dawkins, R. (1998). *Unweaving the rainbow, science, delusion and the appetite for wonder.* Allen Lane.

Dawson, G. (1874). Rainbow and its reflection. *Nature, 9*(226), 322.

de Dominis, M. (1611). *De Radiis Visus Et Lucis*. Venice. https://tinyurl.com/2kk w6nfd.

Descartes, R. (2001). *Discourse on method, optics, geometry and meteorology* (P. J. Olscamp, Trans.). Hackett Publishing Company.

Descartes, R. (1897). *Oeuvres de Descartes: Correspondance I, Avril 1622-Fevrier 1638*. In C. Adam & P. Tannery (Eds.). L. Cerf. https://tinyurl.com/2p8ysyyh.

Descartes, R. (1996). *Discourse on method and meditations on first philosophy*. In D. Weisman (Ed.). Yale University Press.

Deutscher, G. (2010). *Through the language glass*. Heinemann.

Dod, P. (1735). An account of an extraordinary effect of lightning in communicating magnetism. Communicated By Pierce Dod, M.D. F. R. S. From Dr. Cookson of wakefield in Yorkshire. *Philosophical Transactions of the Royal Society, 39*(437), 74–5. https://tinyurl.com/5av9jvfd.

Drake, S. (1980). *Galileo*. OUP.

Edinburgh Review. (1802–1929). https://tinyurl.com/497bkz69.

Eliade, M. (1964). *Shamanism: Archaic techniques of ecstasy* (W. R. Trask, Trans.). Penguin.

Ellis, G. E. (1871). *Memoir of Sir Benjamin Thompson, Count Rumford, with notices of his daughter*. https://tinyurl.com/kthaeykt.

Fara, P. (2002). *Newton*. Macmillan.

Faraday, M. (1822). On some new electro-magnetical motions, and on the theory of magnetism. *Quarterly Journal of Science, Literature, and the Arts, 12*, 74–96. https://tinyurl.com/yuecay2a.

Faraday, M. (1846). Thoughts on ray vibrations. *The London, Edinburgh and Dublin Philosophical Magazine and Journal of Science, 27*(185), 345–350. https://tinyurl.com/457j8hjr.

Faraday, M. (1899). The letters of Faraday and Schenbein, 1836–62, with notes and references to contemporary letters. In G.W.A. Kahlbaum & F.V. Darbishire (Eds.). Basle & London. https://tinyurl.com/yc7tnyfz.

Faraday, M. (1932). *Faraday's diary. Being the various philosophical notes of experimental investigation made by Michael Faraday, during the years 1820–1862 and bequeathed by him to the Royal Institution of Great Britain* (vol. 1). G. Bell and Sons, Ltd. https://tinyurl.com/2anjr4au.

Faraday, M. (1908). *The chemical history of a candle*. London: Chatto & Windus. https://tinyurl.com/224e5za7.

Faraday, M. (1991). *Curiosity perfectly satisfied—Faraday's travels in Europe, 1813–1815*. In B. Bowers & L. Symons (Eds.). Peregrinus in association with The Science Museum.

Feingold, M., & Svorenčík, A. (2020). A preliminary census of copies of the first edition of Newton's principia (1687). *Annals of Science, 77*(3), 253–348.

Finkel, I. (2014). *The arc before Noah, decoding the story of the flood*. Hodder & Stoughton.

Franklin, B. (1941). *Benjamin Franklin's experiments. A new edition of Franklin's experiments and observations on electricity.* In I. Bernard Cohen (Ed.). Harvard University Press. https://tinyurl.com/33u78v9b.

Fresnel, A. (1818). Mémoire sur la diffraction de les Lumière. *Mémoires de l'Académie des Sciences, 5.* https://tinyurl.com/yvkuhunf.

Fresnel, A. (1866). *Oeuvres Completes D'Augustin Fresnel* (vol. 1). Paris: Imprimerie Impériale. https://tinyurl.com/yyszsjvx.

Galileo, G. (1954). *Dialogues concerning two new sciences* (H. Crew & A. de Salvio, Trans.). Dover Publications. https://tinyurl.com/nuuh7dkx.

Galvani, L. (1791). De viribus electricitatis in motus muscularis commentarius. *De Bononiensi Scientiarum et Artium Instituto atque Academia Commentarii, Tomus 7.* https://tinyurl.com/2vmdyz22.

Gamow, G. (1993). *Mr Tompkins in paperback.* CUP.

Gaukroger, S. (1995). *Descartes, an intellectual biography.* OUP.

Gautherot, N. (1801). Mémoire sur le Galvanisme, lu à la classe des sciences de l'Institute, le 26 ventóse an 9. *Annales de Chimie, 39,* 203–219. https://tinyurl.com/2p8zy98h.

George, A. (1999). *The epic of Gilgamesh.* Penguin.

George, A. J. (1938). The genesis of the Académie des sciences. *Annals of Science, 3*(4), 372–401.

Gladstone, W. E. (1858). *Studies on Homer and the Homeric age* (vol. 3). Oxford. https://tinyurl.com/34vkjas6.

Gould, S. J. (1993). *Eight little Piggies: Reflections in natural history.* W. W. Norton & Company.

Grayling, A. C. (2005). *Descartes: The life of René Descartes and its place in his times.* Simon & Schuster.

Grayling, A. C. (2016). *The age of genius: The seventeenth century and the birth of the modern mind.* Bloomsbury.

Greenler, R. (1980). *Rainbows, Halos and Glories.* CUP.

Grimaldi, F. M. (1665). *Physicomathesis de lumine, coloribus et iride.* Bologna. https://tinyurl.com/yckec4ay.

Guerlac, H. (1965). The word spectrum: A lexicographic note with a query. *Isis, 56*(2), 206–207. https://tinyurl.com/36tzb42k.

Halley, E. (1698). An account of the appearance of an extraordinary iris seen at Chester, in August last. *Philosophical Transactions of the Royal Society, 20*(240), 193–196. https://tinyurl.com/2n5frs4z.

Halley, E. (1700). De Iride, Sive de Arcu Caelesti, Differtatio Geometrica, qua Methodo Directa Iridis Ntriusq; Diameter, Data Ratione Refractionis, Obtinetur: Cum Solutione Inversi Problematis, Sive Inventione Rationis Istius ex Data Arcus Diametro. *Philosophical Transactions of the Royal Society, 20*(267), 714–725. https://tinyurl.com/yc6sz8t4.

Hammer, D. (1903). Airy's theory of the rainbow. *Journal of the Franklin Institute, 156*(5), 335–349.

Harriot, T. (1590). *A brief and true report of the new found land of Virginia*. https://tinyurl.com/mu6wb4ry.

Haußmann, A. (2016). Rainbows in nature: Recent advances in observation and theory. *European Journal of Physics, 37*(6), 1–30.

Haydon, B. R. (1926). The autobiography and memoirs of Benjamin Robert Haydon (1786–1846), Edited from his Journals By Tom Taylor. New York: Harcourt Brace and Co. https://tinyurl.com/5n8u9fy4.

Helmholtz, H. (1962). *Treatise on physiological optics* (vol. 2). In J.P.C. Southall (Ed.). Dover Publications.

Helmholtz, H. (1895). *Sensations of tone as a physiological basis for the theory of music* (A. J. Ellis, Trans.). Longmans, Green, and Co.

Herschel, W. (1800a). Investigations of the powers of the prismatic colours to heat and illuminate objects; with remarks, that prove the different refrangibility of radiant heat. To which is added, an inquiry into the method of viewing the sun advantageously, with telescopes of large apertures and high magnifying powers. *Philosophical Transactions of the Royal Society, 90*, 255–283. https://tinyurl.com/3pcphykf.

Herschel, W. (1800b). Experiments on the refrangibility of the invisible rays of the sun. *Philosophical Transactions of the Royal Society, 90*, 284–292. https://tinyurl.com/5bde6rha.

Herschel, W. (1800c). Experiments on the solar, and on the terrestrial rays that occasion heat; with a comparative view of the laws to which light and heat, or rather the rays which occasion them, are subject, in order to determine whether they are the same, or different, Part II. *Philosophical Transactions of the Royal Society, 90*, 437–538. https://tinyurl.com/2p8w6s2h.

Hertz, H. (1888). Über die Ausbreitungsgeschwindigkeit der elektrodynamischen Wirkung. *Annalen Der Physik, 34*, 551–569.

Hesiod. (1913). *Hesiod, Homeric Hymns, Epic Cycle, Homerica* (H. G. Evelyn-White, Trans.). Harvard University Press. https://tinyurl.com/yttstzb9.

Hilts, V. L. (1978). Thomas Young's "Autobiographical sketch." *Proceedings of the American Philosophical Society, 122*(4), 248–260.

Homer. (1950). *The Illiad* (E. V. Rieu, Trans.). Penguin.

Hooke, R. (1672). *Considerations upon Mr. Newton's discourse on light and colours.* In T. Birch (1757). https://tinyurl.com/mwbnsveb.

Hooke, R. (2003). Micrographia: Or some physiological descriptions of minute bodies made by magnifying glasses with observations and inquiries thereupon. Dover Phoenix Editions. https://tinyurl.com/fd2ezb7a.

Hooykaas, J. (1956). The rainbow in ancient Indonesian religion. *Journal of the Humanities and Social Sciences of Southeast Asia, 112*(3), 291–322. https://tinyurl.com/m4xmm9.

Hopkins, G. M. (1948). *Poems of Gerard Manley Hopkins*. In R. Robert Bridges (Ed.). OUP. https://tinyurl.com/4pnzs67p.

Hudson, W. H. (1984). *Idle days in Patagonia*. Dent.

Huff, T. E. (1993). The rise of early modern science. CUP.

Hulst, H. C. van de. (1957). Light scattering by small particles. Dover Books.

Humphreys, W. J. (1938). Why we seldom see Lunar Bows. *Science 88*(2291), 496–498. https://tinyurl.com/krwepetv.

Humphreys, W. J. (1929*). Physics of the air*. McGraw-Hill Book Company. https://tinyurl.com/mtr9nt6x.

Hunter, M. (1984). A 'College' for the royal society: The abortive plan of 1667–1668. *Notes and Records: Royal Society of London, 38*(2), 159–186. https://tinyurl.com/52nr4fyp.

Huygens, C. (1659). *Systema Saturnium*. Hagae-Comitis: Ex typographia Adriani Vlacq. https://tinyurl.com/2x6wh8ss.

Huygens, C. (1673a). *Horologium oscillatorium sive de motu pendularium*. https://tinyurl.com/3u37jwns.

Huygens, C. (1673b). An extract of a letter lately written by an ingenious person from Paris, containing some Considerations upon Mr. Newtons Doctrine of Colors, as also upon the effects of the different Refractions of the Rays in Telescopical Glasses. *Philosophical Transactions of the Royal Society, 8*(96), 6086–6087. https://tinyurl.com/32n9nna3.

Huygens, C. (1690). *Traité de la Lumière*. Leide. https://tinyurl.com/yeyahcam.

Huygens, C. (1895). *Oeuvres Complètes de Christian Huygens* (vol. 6). Société Hollandaise Des Sciences. https://tinyurl.com/3eeb2ysm.

Huygens, C. (1897). *Oeuvres Complètes de Christian Huygens* (vol. 7). Société Hollandaise Des Sciences. https://tinyurl.com/yn5w84bj.

Huygens, C. (1899). *Oeuvres Complètes de Christian Huygens* (vol. 8). Société Hollandaise Des Sciences. https://tinyurl.com/3r5pn6y5.

Huygens, C. (1901). *Oeuvres Complètes de Christian Huygens* (vol. 9). Société Hollandaise Des Sciences. https://tinyurl.com/y38ekm92.

Huygens, C. (1905). *Oeuvres Complètes de Christian Huygens* (vol. 10). Société Hollandaise Des Sciences. https://tinyurl.com/yc823yss.

Huygens, C. (1916). *Oeuvres Complètes de Christian Huygens* (vol. 13). Société Hollandaise Des Sciences. https://tinyurl.com/5djfvw3b.

Huygens, C. (1932). *Oeuvres Complètes de Christian Huygens* (vol. 17). Société Hollandaise Des Sciences. https://tinyurl.com/mr2zc9hb.

Huygens, C. (1912). *Treatise on light* (S. P. Thompson, Trans.). Macmillan and Co. https://tinyurl.com/ywpnj7sv.

James, E. O. (1960). *The ancient gods*. Weidenfield & Nicholson.

James, F. A. J. L. (2011). *Correspondence of Michael Faraday* (vol. 6): 1860–1867. Institution of Engineering and Technology.

Journal Book of the Royal Society 1672–1677. https://tinyurl.com/3d5b9xwh.

Keats, J. (1909). Keats: Poems published In 1820. In M. Robinson (Ed.). At The Clarendon Press. https://tinyurl.com/3tv3dftt.

Keats, J. (1952). The letters of John Keats, 4th edn. In M.B. Buxton Forman (Ed.). OUP. https://tinyurl.com/2ck57p4y.

Keller, A. (1983). *The infancy of atomic physics, Hercules in his cradle*. Clarendon Press.

Kemp, M. (1990). *The science of art*. Yale University Press.

Kepler, J. (1859). *Opera omnia* (vol. 2). In C. Frisch (Ed.). Frankofurti A.M. ET Erlanger, Heyder & Zimmer. https://tinyurl.com/bdfxxtct.

Kepler, J. (2000). *Paralipomena to Witelo and optical part of astronomy* (W. H. Donahue, Trans.). Green Lion Press.

Keynes, J. M. (1947). *Newton the man*. The royal society Newton tercentenary celebrations, 15–19 July 1946. Cambridge University Press.

Können, G. P. (1985). *Polarized light in nature*. CUP. https://tinyurl.com/2p8 ju6dm.

Kuschel, R., & Monberg, T. (1974). "We don't talk much about colour here": A study of colour semantics on Bellona Island. *MAN New Series, 9*(2), 213–242. https://tinyurl.com/3snfh49y.

Langwith, B. (1723). Extracts of several letters to the Publisher, from the Reverend Dr. Langwith, Rector of Petworth in Sussex, concerning the appearance of several arches of colours contiguous to the inner edge of the common rainbow. *Philosophical Transactions of the Royal Society, 32*(375), 241–245. https://tinyurl.com/ypww5v8k.

Laplace, P.-S. (1798–1825). *Traité de Méchanique Celeste*, 5 vols.

Lavoisier, A. L. (1789). *Traité Élémentaire de Chimie*, 3 vols.

Lee, R. L., & Fraser, A. B. (2001). *The rainbow bridge, rainbows in art*. The Pennsylvania State University Press.

Leonardo da Vinci. (1956). *Treatise on painting: [Codex Urbinas Latinus 1270]* (A. Philip McMahon, Trans.; with an introduction by Heydenreich, L.H.). Princeton University Press.

Leslie, C. R. (1912). *Memoirs of the life of John Constable, R.A.* J. M. Dent & Sons Ltd. https://tinyurl.com/3jmmfw8a.

Leslie, C. R. (1896). *Life and letters of John Constable, R.A.* Chapman and Hall. https://tinyurl.com/wu6bcazf.

Lévi-Strauss, C. (1970). The raw and the cooked, introduction to a science of mythology (vol. 1) (J. D. Weightman, Trans.). Harper & Row.

Lewis, G. (1926). Letter to the editor. *Nature, 118*, 874–875.

Lindberg, D. C. (1966). Roger Bacon's theory of the rainbow: Progress or regress? *Isis, 57*(2), 235–248. https://tinyurl.com/yckffzr.

Locke, J. (2007). *John Locke, selected correspondence*. In M. Goldie (Ed.). OUP.

Lodge, O. (1907). Modern views of electricity, 3rd edn. Macmillan and Co, Limited. https://tinyurl.com/9yr4wkec.

Loewenstein, P. J. (1961). Rainbow and serpent. *Anthropos 5*, (1/2), 31–40. https://tinyurl.com/rprwd9f6.

Lovell, D. J. (1968). Herschel's dilemma in the interpretation of thermal radiation. *Isis, 59*(1), 46–60. https://tinyurl.com/26d6pe9v.

Lucretius. (2007). *The nature of things* (A.E. Stallings, Trans.). Penguin.

MacAdam, D. L. (1970). *Sources of color science*. MIT.

MacCormack, S. (1988). Pachacuti: Miracles, punishments, and last judgment: Visionary past and prophetic future in early colonial Peru. *The American Historical Review, 93*(4), 960–1006. https://tinyurl.com/5n6mmzsv.

Magie, W. M. (Ed.). (1963). *A source book in physics.* Harvard University Press. https://tinyurl.com/y92j6t8v.

Mahon, B. (2003). *The man who changed everything.* John Wiley & Sons.

Malebranche, N. (1978). *Oeuvres completes, correspondance et actes, 1690–1715* (vol. 19). Librairie Philosophique Vrin.

Mariotte, E. (1717). *Oeuvres De Mariotte* (vol. 1). Marchand Libraire, Imprimeur de l'Université de la Ville. https://tinyurl.com/z7x9hzb7.

Maxwell, J. C. (1862). On physical lines of force, part III: The theory of molecular vortices applied to statical electricity. *The London, Edinburgh, and Dublin Philosophical Magazine and Journal of Science, 23*(151), 12–24.

Maxwell, J. C. (1865). A dynamical theory of the electromagnetic field. *Philosophical Transactions of the Royal Society, 155*, 459–512. https://tinyurl.com/3mutcvyd.

Miller, W. H. (1841). On spurious rainbows. *Transactions of the Cambridge Philosophical Society, 7*(3), 277–286. https://tinyurl.com/2p86z452.

Minnaert, M. (1954). *The nature of light & colour in the open air.* Dover Books.

Montgomery, S. L. (1999). *The moon and the western imagination.* The University of Arizona Press.

Murdoch, I. (1998). *Existentialists and mystics.* Penguin.

Naylor, J. (2002). *Out of the Blue.* CUP.

Newton, H. (1727/8). *Two letters from Humphrey Newton to John Conduitt.* Keynes Ms. 135, King's College Cambridge https://tinyurl.com/2sunwxzy.

Newton, I. (1668/9). *Letter from Newton to a friend, together with Collins's description of a telescope mentioned in the Newton letter.* https://tinyurl.com/2p8hac7t.

Newton, I. (1671a/2). *A letter to Mr Oldenburg containing an advise about the Metalline composition for Mr Newtons reflecting telescope.* https://tinyurl.com/dtdbrd75.

Newton, I. (1671b/2). A letter of Mr. Isaac Newton, Professor of the mathematicks in the University of Cambridge; containing his New theory about light and colors. *Philosophical Transactions of the Royal Society, 6*(80), 3075–3087. https://tinyurl.com/bdnktcjr.

Newton, I. (1672). Mr Isaac Newtons answer to some considerations upon his doctrine of light and colors; which doctrine was printed in numb. 80. of these tracts. *Philosophical Transactions of the Royal Society, 7*(88), 5084–5103. https://tinyurl.com/3h2et45z.

Newton, I. (1673). Mr. Newtons answer to the foregoing letter further explaining his theory of light and colors, and particularly that of Whiteness; together with his continued hopes of perfecting telescopes by reflections rather than refractions. *Philosophical Transactions of the Royal Society, 8*(96), 6087–6092. https://tinyurl.com/3fkbksje.

Newton, I. (1675a). *An hypothesis explaining the properties of light.* In T. Birch (1757) (Ed.) (pp. 247–305). https://tinyurl.com/2p9eefv3.

Newton, I. (1675b). Letter from Newton to Henry Oldenburg, dated 21 Dec, 1675b. In H.W. Turnbull (Ed.) (1959) (p. 406). https://tinyurl.com/y3yrmp54.

Newton, I. (1686). Letter to Edmund Halley on the doctrine of projectiles and motions of the heavens 20th June 1686. https://tinyurl.com/yyrt2a2x.

Newton, I. (1692a). Letter from Isaac Newton to Richard Bentley, dated 10 December, 1692a. In I. Bernard Cohen & R.S. Westfall (Eds.) (1995). https://tinyurl.com/3mxt5buh.

Newton, I. (1692b/3). Letter from Isaac Newton to Richard Bentley, 25 February 1692b/3. In I. Bernard Cohen & R.S. Westfall (Eds.) (1995). https://tinyurl.com/3n48znfd.

Newton, I. (1952). *Opticks, or a treatise on the reflections, refractions, inflections & colours of light*. Dover Publications. https://tinyurl.com/mrxe85ub.

Newton, I. (1960). *Sir Isaac Newton's mathematical principles of natural philosophy and his system of the world* (F. Cajori, Trans.). University of California Press. https://tinyurl.com/seaxjwvt.

Newton, I. (1984). *The optical papers of Isaac Newton* (vol. 1). *The optical lectures 1670–1672*. In A.E. Shapiro (Ed.). CUP.

Nietzsche, F. (1967). The will to power, a new translation (W. Kaufmann & R. J. Hollingdale, Trans.). New York: Random House.

Niven, W. D. (Ed.). (1890). *The scientific papers of James Clerk Maxwell* (vol. 1 & 2). CUP. https://tinyurl.com/9bskz8nr, https://tinyurl.com/bddmwd7t.

Nollet, J. A. (1746). Observations sur quelques nouveaux phénomènes d'Électricité. *Mémoires de l'Academie Royale des Science*. https://tinyurl.com/2ssehzt9.

Ockenden, R. E. (1936). Marco Antonio de Dominis and his explanation of the rainbow. *Isis, 26*(1), 40–49.

Oersted, H. C. (1820). Experiments on the effect of a current of electricity on the magnetic needle. *Annals of Philosophy, or, Magazine of Chemistry, XVI*. https://tinyurl.com/2p8xycjz.

Oestigaard, T. (2019). *Rainbows, pythons and waterfalls, Heritage, poverty and sacrifice among the Busoga, Uganda*. The Nordic Africa Institute. https://tinyurl.com/26arf2fx.

Osler, M. J. (2006). A hero for their times: Early biographies of Newton. *Notes and Records: Royal Society of London, 60*, 291–305. https://tinyurl.com/4rj4pfpx.

Ovid. (1986). *Metamorphoses* (A. D. Melville, Trans.). OUP.

Palmer, F. (1945). Unusual rainbows. *American Journal of Physics, 13*(3), 203–204.

Paris, J. A. (1831). *The life of Sir Humphry Davy* (vol. 1). Henry Colburn and Richard Bentley. https://tinyurl.com/4jur9u4n.

Peacock, G. (1855). *Life of Thomas Young, M.D., F.R.S.* John Murray. https://tinyurl.com/2r8rw8w2.

Pemberton, H. (1723). A letter to Dr. Jurin, Coll. Med. Lond. Soc. & Secr. R. S. concerning the abovementioned appearance in the rainbow, with some other reflections on the same subject. *Philosophical Transactions of the Royal Society, 32*(375), 245–261. https://tinyurl.com/5f88w7jv.

Pernter, J. M., & Exner, F. M. (1922). *Meteorologische Optik*. Wilhelm Braumüller, Wien und Leipzig.

Plato. (1952). *Theaetetus* (H. N. Fowler, Trans.). Harvard University Press.

Potter, R. (1838). Mathematical considerations on the problem of the rainbow, shewing it to belong to physical optics. *Transactions of the Cambridge Philosophical Society, 6*, 141–152.

Potter, R. (1856). *Physical optics*. Walton And Maberly. https://tinyurl.com/4kx uy9w3.

Preston, T. (1912). The theory of light. Macmillan and Co. https://tinyurl.com/4h5 bfyep.

Rashed, R. (1990). A pioneer in anaclastics: Ibn Sahl on burning mirrors and lenses. *Isis, 81*(3), 464–491. https://tinyurl.com/mryz7h3a.

Robertson, R. (2020). *The enlightenment, the pursuit of happiness, 1680–1790*. Harper Collins.

Rømer, O. (1676). Demonstration Touchant Le Mouvement De La Lumiere Trouvé Par M. Römer. *Journal des Scavans, Decembre*, 233–236. https://tinyurl.com/bdh 52h7x.

Ronchi, V. (1970). *The nature of light* (V. Barocas, Trans.). Heinemann.

Röentgen, W. K. (1895). Ueber eine neue Art von Strahlen. Vorläufige Mitteilung. Sitzungsberichten der Würzburger Physikalischen-Medicinischen Gesellschaft.

RSV Bible. (1952). https://tinyurl.com/yc78c9wy.

Rupert Hall, A. (1990). Beyond the Fringe: Diffraction as seen by Grimaldi, Fabri, Hooke and Newton. *Notes and Records: Royal Society of London, 44*(1), 13–23. https://tinyurl.com/2p8ae9s6.

Rupert Hall, A. (1992). *Isaac Newton, adventurer in thought*. Blackwell.

Rupert Hall, A., & Boas Hall, M. (1962). *Unpublished scientific papers of Isaac Newton, a selection from the Portsmouth collection in the university library*. CUP.

Ruskin, J. (1906). *Modern painters* (vol. 3). J.M.Dent & Co. https://tinyurl.com/ 55tfpb37.

Sabra, A. I. (1981). *Theories of light from Descartes to Newton*. CUP.

Sarton, G. (1928). The foundation of electromagnetism (1820). *Isis, 10*(2), 435–444. https://tinyurl.com/mr4bha66.

Sarton, G. (1931). The discovery of the electric cell (1800). *Isis, 15*(1), 124–157. https://tinyurl.com/2p8yauvn.

Sayili, A. M. (1939). The Aristotelian explanation of the rainbow. *Isis, 30*(1), 65–83. https://tinyurl.com/2p9exzzy.

Scott, W. (1891). *The Journal of Sir Walter Scott, From the Original Manuscript at Abbotsford* (vol. 1). Harper Bros, Franklin Square. https://tinyurl.com/487sadce.

Shapiro, A. E. (1980). The evolving structure of Newton's theory of white light and color. *Isis, 71*(2), 211–235. https://tinyurl.com/2p967d6n.

Shapiro, A. E. (Ed.). (1984). *The optical papers of Isaac Newton* (vol. 1). CUP.

Shapiro, A. E. (1989). Huygens' Traité De La Lumière and Newton's Opticks: Pursuing and Eschewing Hypotheses. *Notes and Records: Royal Society of London, 43*(2), 223–247. https://tinyurl.com/mub79t5h.

Shapley, D. (1949). Pre-Huygenian observations of saturn's ring. *Isis, 40*(1), 12–17. https://tinyurl.com/ytpy552r.

Shea, J. H. (1998). Ole Rømer, the speed of light, the apparent period of Io, the Doppler effect, and the dynamics of Earth and Jupiter. *American Journal of Physics, 66*(7), 561–568.

Shea, W. R. (1991). *The magic of numbers and motion*. Watson Publishing International.

Sheehan, W. (1988). *Planets & perception*. University of Arizona Press.

Shepherd, N. (2011). *The living mountain*. Canongate.

Simpson, T. K. (1966). Maxwell and the direct experimental test of his electromagnetic theory. *Isis, 57*(4), 411–432. https://tinyurl.com/f69nr2ub.

Snyder, L. J. (2011). *The philosophical breakfast club, four remarkable friends who transformed science and changed the world*. Broadway Paperbacks.

Sobel, D. (1998). *Longitude: The true story of a lone genius who solved the greatest scientific problem of his time*. Fourth Estate.

Somerville, M. (1831). *Mechanism of the heavens*. John Murray. https://tinyurl.com/yfevu8bb.

Stachel, J. (Ed.). (1998). *Einstein's miraculous year: Five papers that changed the face of physics*. Princeton University Press.

Steffens, H. J. (1977). *The development of Newtonian optics in England*. Science History Publications.

Stroup, A. (1987). Royal funding of the Parisian Académie royale des sciences during the 1690s. *Transactions of the American Philosophical Society, 77*(4), 1–167. https://tinyurl.com/2jy95k7v.

Tait, P. G. (1880). Clerk-Maxwell's scientific work. *Nature, 21*, 317–321.

Tait, P. G. (1898). *Scientific papers* (vol. 1). Cambridge University Press. https://tinyurl.com/y3ykyr9m.

Tape, W., Seidenfaden, E., & Können, G. P. (2008). The legendary Rome halo displays. *Applied Optics, 47*(34), 72–84.

Tarasov, L. V., & Tarasova, A. N. (1984). *Discussions on the refraction of light*. Mir Publishers. https://tinyurl.com/2p8b5tf7.

Theusner, M. (2011). Photographic observation of a natural fourth-order rainbow. *Applied Optics, 50*(28), 129–133.

Thomas, J. M. (1999). Sir Benjamin Thompson, count Rumford and the royal institution. *Notes and Records: Royal Society of London, 53*(1), 11–25. https://tinyurl.com/bdtazv7d.

Thomson, J. (1908). *Complete poetical works of James Thomson*. In Robertson, J.L. (Ed). OUP. https://tinyurl.com/55barw99.

Thornes, J. E. (1999). *John Constable's skies*. University of Birmingham Press.

Tricker, R. A. R. (1970). *Introduction to meteorological optics*. Mills & Boon.

Turnbull, W. H. (Ed.). (1959). *The correspondence of Isaac Newton (1661–1675)*. CUP.

Turnbull, W. H. (Ed.). (1960). *The correspondence of Isaac Newton (1676–1687)*. CUP.

Turnbull, W. H. (Ed.). (1961). *The correspondence of Isaac Newton (1688–1694)*. CUP.

Turner, L. A. (1993). The rainbow as the sign of the covenant in genesis IX 11–13. *Vetus Testamentum, 43*(1), 119–124.

Twain, M. (1872). *Roughing it*. American Publishing Company. https://tinyurl.com/ryrwkd5j.

Tyndall, J. (1868). *Faraday as a discoverer*. Longmans, Green & Co. https://tinyurl.com/mpjhhbcd.

Tyndall, J. (1884). On rainbows. *The London, Edinburgh And Dublin Philosophical Magazine And Journal Of Science, 17*(103), 61–64. https://tinyurl.com/mub2u24j.

Urton, G. (1981). Animals and astronomy in the Quechua universe. *Proceedings of the American Philosophical Society, 125*(2), 110–127. https://tinyurl.com/ppz4d88t.

Van Helden, A. (1974). The telescope in the 17th century. *Isis, 65*(1), 38–58. https://tinyurl.com/3ducyb39.

Venturi G. B. (1814). *Commentari sopra la storia e le teorie dell'ottica*. Bologna. https://tinyurl.com/ybp2j45m.

Virgil. (1697). *Aeneid*. https://tinyurl.com/2nh3z2aw.

Volta, A. (1800). On the electricity excited by the mere contact of conducting substances of different kinds. *Philosophical Transactions of the Royal Society, 90*, 403–431 https://tinyurl.com/2p8v8jws.

Voltaire. (1980). *Letters on England* (L. Tancock, Trans.). Penguin Books.

Voltaire. (1738). *Eleméns de la philosophie de Neuton*. A Londres. https://tinyurl.com/a6rbnsz4.

Walker, J. (1977). *How to create and observe a Dozen rainbows in a single drop of water*. Scientific American. https://tinyurl.com/3k5v82sd.

Weinberg, S. (2015). *To explain the world, the discovery of modern science*. Allen Lane.

Werrett, S. (2001). Wonders never cease: Descartes's météores and the rainbow fountain. *BJHS, 34*(2), 129–147. https://tinyurl.com/yrncr5zj.

Wessing, R. (2006). Symbolic animals in the land between the waters: Markers of place and transition. *Asian Folklore Studies, 65*(2), 205–239. https://tinyurl.com/msn7w6ns.

Westfall, R. S. (1980). *Never at rest*. CUP.

Wheaton, B. R. (1978). Philipp Lenard and the photoelectric effect, 1889–1911. *Historical Studies in the Physical Sciences, 9*: 299–322. https://tinyurl.com/2ncsk3ck.

Wills, G. H. A. (1937). *Untitled letter, 15/10/1937*. The Times.

Williams, L. P. (1965). *Michael Faraday: A biography*. Chapman & Hall.

Williams, L. P. (1983). What were Ampère's earliest discoveries in electrodynamics? *Isis, 74*(4), 92–508. https://tinyurl.com/2bcprw4v.

Wollaston, W. H. (1802). On the oblique refraction of Iceland crystal. *Philosophical Transactions of the Royal Society, 92*, 381–386. https://tinyurl.com/bdhf8c74.

Xenophanes of Colophon. (1992). *Xenophanes of colophon fragments: A text and translation with a commentary* (J. H. Leshe, Trans.). University of Toronto Press.

Young, T. (1802a). On the theory of light and colours. Philosophical *Transactions of the Royal Society, 92*, 12–48. https://tinyurl.com/ye256c9x.

Young, T. (1802b). An account of some cases of the production of colours, not hitherto described. *Philosophical Transactions of the Royal Society, 92*, 387–97. https://tinyurl.com/3d4h632v.

Young, T. (1804). Experiments and calculations relative to physical optics. *Philosophical Transactions of the Royal Society, 94*, 1–16. https://tinyurl.com/24923h3k.

Young, T. (1845). *A course of lectures on natural philosophy and the mechanical arts* (vol. 1). Taylor & Walton, London. https://tinyurl.com/5h29faaw.

Young, T. (1855). *Miscellaneous works of the late Thomas Young M.D. F.R.S.* (vol. 1). In G. Peacock (Ed.). John Murray. https://tinyurl.com/3b9yx3e9.

Zhao, Q. (1992). *A study of dragons, east and west.* Peter Lang.

Ziggelaar, A. (1980). How did the wave theory of light take shape in the mind of Christiaan Huygens? *Annals of Science, 34*(2), 179–187.

Zinsser, J. P. (2001). Translating Newton's principia: The Marquise Du Châtelet's revisions and additions for a French audience. Notes and Records: *Royal Society of London, 55*(2), 227–245. https://tinyurl.com/4zta6wtj.

Index

Printed in the United States
by Baker & Taylor Publisher Services